Design of Liquid-Retaining Concrete Structures

To Joy Elizabeth

Design of Liquid-Retaining Concrete Structures

R.D. ANCHOR, B.Sc., C.Eng., F.I.C.E., F.I.Struct.E.

Senior Partner, R.D. Anchor Consultants
Visiting Senior Lecturer in Civil Engineering,
University of Aston in Birmingham, U.K.

A HALSTED PRESS BOOK

John Wiley and Sons
New York—Toronto

Published by Surrey University Press
A member of the Blackie Group
Bishopbriggs, Glasgow and
Furnival House, 14–18 High Holborn, London

Published in the U.S.A. and Canada by
Halsted Press,
a Division of John Wiley and Sons Inc., New York

Library of Congress Cataloging in Publication Data

Anchor, R. D.
    Design of liquid-retaining concrete structures.
    "A Halsted Press book."
    Bibliography: p.
    Includes index
    1. Concrete tanks.   2. Reservoirs.   3. Reinforced
concrete construction.   I. Title.
TA683.2.A57   1981   681'.766   80-29093
ISBN 0-470-27123-X

Printed in Great Britain

Series Foreword

This book forms part of a series dealing with the design of structures. The emphasis is on the word design as this is the crucial part of an engineer's activity, which distinguishes him from a scientist working in the related field. The engineer must be able to synthesize the information available for the purpose of creating on paper that which is to be built, i.e. designing. This activity requires knowledge of analysis of the whole as well as of the components of that which is being designed. It requires also knowledge of the codes and standards in force and what is considered to be good engineering practice. Finally, design requires engineering judgement, and judgement can come only with experience.

This is why a university course, even if design-oriented as at some universities, does not produce a designer. The young graduate, however bright, when confronted with his first, and possibly not only first, design job simply lacks the background for his new task. This is where the books in the present series come in: they make it possible to approach the task of design in a reasonable manner.

The books cover mainly the field of structural design. Recent titles in the series are *Design of Composite Steel-Concrete Structures* (L. C. P. Yam), *Glass-Reinforced Plastics in Construction* (L. Holloway) and *Design of Structural Steelwork* (P. R. Knowles). These are the sorts of design problems with which the civil engineer is concerned. The books explain relatively simply the background to the design problem and then in quite some detail, the main features of the design process. All this is fully illustrated by worked examples. Such an approach may seem old-fashioned to an enthusiast of pure analysis, but example and precept are essential if modern design is to build upon the accumulated stores of successful design used in the past. Indeed, modern disciplines have borrowed our approach but re-named it 'case study'.

The books in the present series should thus prove of great value to the young engineer and also to his senior colleague who is designing in an unfamiliar field. For these people the book is a 'must'. But it also a wise investment (and in these inflationary days such is not easy to come by) for the undergraduate who appreciates the importance of design. With the aid of the book he can profit much more from his university or polytechnic course and enter employment much better prepared.

The authors of the books in the series are all specialists willing, in the best spirit of the engineering profession, to share their knowledge and experience. I am therefore confident that the present series not only fills a genuine need but does so really well.

Adam Neville, General Editor
University of Dundee
May, 1980

Preface

The design of any structure is a complicated matter—particularly so for civil engineering structures where the designer normally acts as structural engineer and 'architect'. Not only does the designer prepare the engineering design of the structure; he also has to consider the general layout, pipework, mechanical and electrical services, and not least, the appearance.

In a book of this size devoted to the elements of design which are applicable to liquid-retaining structures, it is not possible to deal with each separate type of reservoir or tank, and it is certainly not practicable to consider the non-structural details. I have aimed to give a full description of the design of each structural element, so that the reader will be able to design each element of any structure presented to him.

Detailed calculations are given for a range of structural types, and design tables and charts have been included to make the book complete in itself. The notation follows the international system, and metric units are used. The choice of units follows current UK engineering practice, rather than adhering strictly to the S.I. system. The calculations are presented mostly to an accuracy of three significant figures, which is adequate in relation to the accuracy of the basic data and material strength. The author makes no apology for using a value of $10\,kN/m^3$ for the weight of water rather than the more accurate value of 9.81. The difference is less than 2%.

Acknowledgements

I am conscious of the specific and indirect assistance that I have received from my friends and fellow-engineers in writing this book and would like to mention in particular Messrs. A. Allen, A. Astill, A. W. Hill, Professors M. Holmes and B. P. Hughes, and Dr. R. Savidge.

I must also thank Anthony J. Harman who has drawn the figures and calculation sheets with care and enthusiasm. Appendix A, together with the design charts in Appendix B, are reproduced by permission of the Cement and Concrete Association from *Handbook to BS 5337* by R. D. Anchor, A. W. Hill and B. P. Hughes (Viewpoint Publication 14-011). Figure 1.2 is reproduced by kind permission of Thomas Garland and Partners, Consulting Engineers, Dublin.

R. D. Anchor

Contents

Notation

a_a distance between the point considered and the axis of the nearest longitudinal bar

a_{cr} distance between the point considered and the surface of the nearest longitudinal bar

a' distance between the compression face and the point at which the crack width is being calculated

A_s area of steel reinforcement

A_{sv} cross-sectional area of shear reinforcement

b breadth (or width) of section

b_t width of the section at the centroid of the tension steel

c nominal cover to tension steel

c_{min} minimum cover to the tension steel

d effective depth of tension reinforcement; diameter of tank

E_c modulus of elasticity of concrete

E_s modulus of elasticity of steel reinforcement

f_b average bond strength between concrete and steel

$f_{c'}$ characteristic cylinder strength of concrete at 28 days

f_{ct} direct tensile strength of the concrete; tensile stress in concrete

f_{cu} characteristic cube strength of concrete at 28 days

f_{dst} design steel stress in tension (allowable stress for limit-state design or permissible stress for alternative design)

f_{dsv} design steel stress in shear reinforcment (i.e. allowable stress for limit-state design or permissible stress for alternative design)

f_k characteristic strength

f_{st} stress in the tension reinforcement

f_t ring tension per unit length

f_y characteristic strength of the reinforcement

F_t	ring tension
h	overall depth of member
H	depth of liquid
h_c	diameter of column or column head
l_1	length of panel in the direction of span, measured from the centres of columns
l_2	width of a panel measured from the centres of columns
l_m	average of l_1 and l_2
L	length, span
m	bending moment per unit length
M	bending moment
M_d	design (service) moment of resistance
M_u	ultimate moment of resistance
n	total load per unit area (CP 110 ultimate load; CP 114 service or working load)
n_b	number of bars in width of section
q	distributed live load per unit length or per unit area
r	radius of tank
s	spacing
s_{max}	estimated maximum crack spacing
s_{min}	estimated minimum crack spacing
t	thickness of wall of tank
T_1	fall in temperature from hydration peak to ambient
T_2	seasonal fall in temperature
v	shear stress; shear force per unit length
v_c	critical concrete shear stress for ultimate limit state
V	total shear force
w_g	unit weight
w_{max}	estimated maximum crack width
x	depth of the neutral axis
z	lever arm
α	coefficient of thermal expansion of mature concrete; coefficient
α_e	modular ratio
β	coefficient
γ_f	partial safety factor for load
γ_m	partial safety factor for material strength
ε_{cs}	estimated shrinkage strain
ε_m	average strain at the level at which cracking is being considered, allowing for the stiffening effect of the concrete in the tension zone (see Appendix A)
ε_{te}	estimated total thermal contraction strain after peak temperature due to heat of hydration

ε_1 strain at the level considered, ignoring the stiffening effect of the concrete in the tension zone

ε_{ult} ultimate concrete tensile strain

θ inclination of shear reinforcement to longitudinal axis of member

ρ steel ratio based on bd; density of liquid

ρ_c steel ratio based on gross concrete section

ρ_{crit} critical steel ratio, based on gross concrete section

ϕ bar size

Metric units

The units of measure used in this book are those which are currently widely used in the United Kingdom. They are based on the metric SI system but are not quite 'pure'. For the benefit of readers who are not used to metric units, an approximate conversion is given below.

Quantity	Unit
Length	metre (m)
Member sizes, etc.	millimetre (mm)
Force	newton (N)
	kilonewton (kN)
Stress, pressure	N/mm^2, kN/mm^2, kN/m^2

Approximate conversions

1 metre	= 39 inches
100 mm	= 4 inches
300 mm	= 1 foot
1 kN	= 2 cwt
$7 N/mm^2$	= 1000 lb/sq. in.
$100 kN/m^2$	= 1 ton/sq. ft.

Note: $1 N/mm^2 = 1 MN/m^2 = 1 MPa$

Concrete strength

The formulae used in this book are based on British practice, where concrete strength is evaluated using test cubes rather than cylinders.

The relation between cube and cylinder strength is usually taken as

$$\frac{f_{c'}}{f_{cu}} = 0.78$$

but the ratio varies widely according to the type of aggregate.

Note

Characteristic loads and strengths are those values used for design purposes and are based on a statistical evaluation.

Service loads and stresses are calculated with applied characteristic loads and generally a partial safety factor for loads equal to 1.0.

Ultimate loads and stresses are calculated with applied characteristic loads and a partial safety factor for loads which generally varies between 1.2 and 1.6.

1 Introduction

1.1 Scope

It is common practice to use reinforced or prestressed concrete structures for the retention, exclusion, or storage of water and other aqueous liquids. Concrete is generally the most economical material of construction and, when correctly designed and constructed, will provide long life and low maintenance costs. The types of structure which are covered by the design methods given in this book are: storage tanks, reservoirs, swimming pools, elevated tanks, ponds, basement walls, and similar structures (figures 1.1 and 1.2). Specifically excluded are: dams, structures subjected to dynamic forces; and pipelines, aqueducts or other types of structure for the conveyance of liquids.

It is convenient to discuss designs for the retention of water, but the principles apply equally to the retention of other aqueous liquids. In particular, sewage tanks are included. The pressures on a structure may have to be calculated using a specific gravity greater than unity, where the stored liquid is of greater density than water. Throughout this book it is assumed that water is the retained liquid unless any other qualification is made.

The design of structures to retain oil, petrol and other penetrating liquids is not included and is dealt with in specialist literature[1]. Likewise, the design of tanks to contain hot liquids is not discussed[2].

1.2 General design objectives

A structure that is designed to retain liquids must fulfil the requirements for normal structures in having adequate strength, durability, and freedom from excessive cracking or deflection. In addition, it must be designed so that the liquid is not allowed to leak or percolate through the concrete structure. In the design of normal building structures, the most critical aspect of the design is to ensure that the structure retains its stability under the imposed loads. In the design of structures to retain liquids, it is usual to find that, if the structure has been proportioned and reinforced so that the liquid is retained without leakage, then the strength is more than adequate. The requirements for ensuring a reasonable service life for the structure without undue maintenance are more onerous for liquid-retaining structures than for normal structures, and adequate concrete cover to the reinforcement is essential. Equally, the concrete itself must be of good quality, and be properly compacted.

Potable water from moorland areas may contain free carbon dioxide or dissolved salts from the gathering grounds which attack normal concrete. Similar difficulties may occur with tanks which are used to store sewage or industrial liquids. After investigating by tests the type of aggressive elements that are present, it may be necessary to increase the cement content of the concrete mix, use special cements or, under severe conditions, use a special lining to the concrete tank[3,4].

1.3   Fundamental design methods

Historically, the design of structural concrete has been based on elastic theory, with specified maximum design stresses in the materials at working loads. More recently, limit state philosophy has been introduced, providing a more logical basis for determining factors of safety. In limit state design, the working or characteristic loads are enhanced by being multiplied by a *partial safety factor*. The enhanced or ultimate loads are then used with the failure strengths of the materials to design the structure. Limit state design methods are now widely used throughout the world for normal structural design[5,6,7].

Formerly, the design of liquid-retaining structures was based on the use of elastic design, with material stresses so low that no flexural tensile cracks developed. This led to the use of thick concrete sections with copious quantities of mild steel reinforcement. The probability of shrinkage and thermal cracking was not dealt with on a satisfactory basis, and nominal quantities of reinforcement were specified in most codes of practice. Over the last few years, analytical procedures have been developed to enable flexural crack widths to be estimated and compared with specified maxima[8]. A method of calculating the effects of thermal and shrinkage strains has also been published[9]. These two developments enable limit state methods to be extended to the design of liquid-retaining structures.

The designer has a choice—he may use either elastic or limit state design. Limit state design is preferable on logical grounds, and will enable some savings in material costs to be made. Elastic design is a simpler process but, when design charts and tables are available, there should be no difficulty in using limit state design.

Figure 1.1   Swimming pool at Altrincham
Architects: Gelsthorpe, Savidge and Simpson
Structural Engineers: R.D. Anchor, Consultants

1.4   Codes of practice

Structural design is often governed by a Code of Practice appropriate to the location of the structure. Whilst the basic design principles are similar in each code, the specified stresses and factors of safety may vary. It is important to consider the climatic conditions at the proposed site, and not to use a code of practice written for temperate zones in parts of the world with more extreme weather conditions.

Two widely used codes are:

1. British Standard Code of Practice BS 5337:1976: The structural use of concrete for retaining aqueous liquids[10].

2. American Concrete Institute ACI 350 R-77: Concrete Sanitary Engineering Structures[11].

Both codes include material specifications, joint details, and design procedures to limit cracking.

BS 5337[10]

British Standard Code of Practice BS 5337:1976 was originally known as CP 2007 and was completely rewritten for the 1976 publication. It includes sections devoted to both limit state design (with references to CP 110) and limiting stress design (referred to as the alternative method of design).

ACI 350 R-77[11]

The original report appeared in 1971 and was amended for the 1977 edition. Structural design is included, with special emphasis on minimizing the possibility of cracking.

1.5   Impermeability

Concrete for liquid-retaining structures must have low permeability. This is necessary to prevent leakage through the concrete and also to provide adequate durability, resistance to frost damage, and protection against corrosion for reinforcement and other embedded steel. An uncracked concrete slab of adequate thickness will be impervious to the flow of liquid if the concrete mix has been properly designed and compacted into position. The specification of suitable concrete mixes is discussed in chapter 2. The minimum thickness of concrete for satisfactory performance in most structures is 200 mm. Thinner slabs should only be used for structural members of very limited dimensions or under very low liquid pressures.

Liquid loss may occur at joints that have been badly designed or constructed, and also at cracks. It is nearly impossible to construct all but the simplest of structures without cracks being present, but if a concrete slab cracks for any reason, there is a possibility that liquid may leak or a damp patch will appear on the face of the slab. It is found however that cracks of limited width do not allow liquid to leak[12] and the problem for the designer is therefore to design the structure so that surface crack widths are limited to a predetermined size.

1.6   Site conditions

The choice of site for a reservoir or tank is usually dictated by requirements outside the structural designer's responsibility, but the soil conditions may

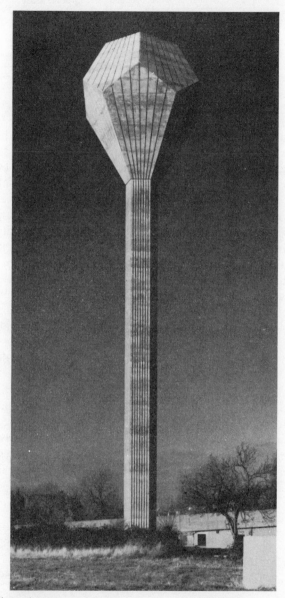

Figure 1.2 Elevated water tower, Dublin
Architect: Andrzej Wejchert in association with Robinson, Keefe and Devane
Structural Engineers: Thomas Garland & Partners, Consulting Engineers

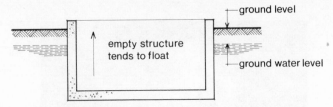

Figure 1.3   Tank flotation due to ground water

radically affect the design. A well-drained site with underlying soils having a uniform safe bearing pressure at foundation level is ideal. These conditions may be achieved for a service reservoir near the top of a hill, but at many sites where sewage tanks are being constructed, the subsoil has a poor bearing capacity and the ground water table is near to the surface. A high level of ground water must be considered in designing the tanks in order to prevent flotation (figure 1.3), and poor bearing capacity may give rise to increased settlement. Where the subsoil strata dip, so that a level excavation intersects more than one type of subsoil, the effects of differential settlement must be considered (figure 1.4). A soil survey is always necessary unless an accurate record of the subsoil is available. 150 mm diameter boreholes should be drilled to a depth of about 10 m, and soil samples taken and tested to determine the sequence of strata and the allowable bearing pressure at various depths. The information from boreholes should be supplemented by digging trial pits with a small excavation to a depth of 3–4 metres.

The soil investigation must also include chemical tests on the soils and ground water to detect the presence of sulphates or other chemicals in the ground which could attack the concrete and eventually cause corrosion of the reinforcement[4]. Careful analysis of the subsoil is particularly important when the site has previously been used for industrial purposes, or where ground water from an adjacent tip may flow through the site. Further information is given in chapter 2.

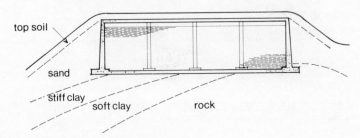

Figure 1.4   Effect on settlement of varying strata

When mining activity is suspected, a further survey may be necessary and a report from the mineral valuer or a mining consultant is necessary. Deeper, randomly located boreholes may be required to detect any voids underlying the site. The design of a reservoir to accept ground movement due to future mining activity requires the provision of extra movement joints and is outside the scope of this textbook[13].

1.7 Influence of construction methods

Any structural design has to take account of the constructional problems involved and, in the field of liquid-retaining structures, this is particularly the case. Construction joints in building structures are not normally shown on detailed drawings but are described in the specification. For liquid-retaining structures, construction joints must be located on drawings, and the contractor is required to construct the works so that concrete is placed in one operation between the specified joint positions. The treatment of the joints must be specified, and any permanent movement joints must be fully detailed. All movement joints require a form of waterstop to prevent leakage, but the designer should specify when waterstops are required at construction joints, and what type. Details of joint construction are given in chapter 5. In the author's opinion, the detailed design and specification of joints is the responsibility of the designer and not the contractor. The quantity of distribution reinforcement and the spacing of joints are interdependent. Casting one section of concrete adjacent to another section, previously cast and hardened, causes restraining forces to be developed which tend to cause cracks in the newly placed concrete. It follows that the quantity of distribution reinforcement also depends on the degree of restraint provided by the adjacent panels.

Any tank which is to be constructed in water-bearing ground must be designed so that the ground water can be excluded during construction. The two main methods of achieving this are by general ground de-watering, or by using sheet piling. If sheet piling is to be used, consideration must be given to the positions of any props that are necessary, and the sequence of construction which the designer envisages[14].

1.8 Design procedure

As with many structural design problems, it is relatively simple to analyse the strength of a structural member, and to calculate the crack widths under load, once the member size and reinforcement have been defined; but the designer has to estimate the size of the members that he proposes to use before any calculations can proceed. With liquid-retaining structures, crack-width calculations control the thickness of the member, and therefore it is impossible to estimate the required thickness directly unless the limiting stress method of design is used. Because of these difficulties, design charts have been prepared which allow designers to choose directly the section thickness and quantity of reinforcement required. The design charts enable limit state designs to be prepared quickly and accurately (see Appendix A).

An intermediate method of design is also possible where the limit state of cracking is satisfied by limiting the reinforcement stress rather than by preparing a full calculation. This procedure is particularly useful for sections under combined flexural and direct stresses. Figure 1.5 illustrates the options available to the designer.

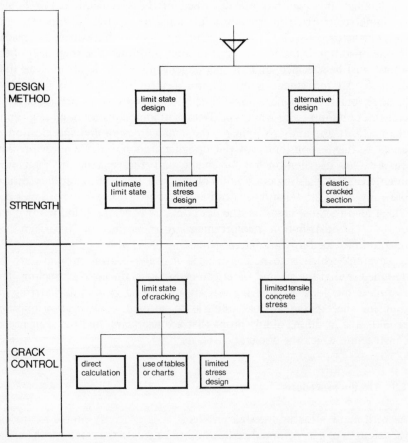

Figure 1.5 Design methods

2 Basis of Design

2.1 Structural action

It is necessary to start a design by deciding on the type and layout of structure to be used. Tentative sizes must be allocated to each structural element, so that an analysis may be made and the sizes confirmed.

All liquid-retaining structures are required to resist horizontal forces due to the liquid pressures. Fundamentally there are two ways in which the pressures can be contained:

(a) By forces of direct tension or compression (figure 2.1)
(b) By flexural resistance (figure 2.2)

Structures designed using tensile or compressive forces are normally circular and may be prestressed. Rectangular tanks or reservoirs rely on flexural action using cantilever walls, propped cantilever walls, or walls spanning in two directions. A structural element acting in flexure to resist liquid pressure reacts on the supporting elements and causes direct forces to occur. The simplest illustration (figure 2.3) is a small tank. Additional reinforcement is necessary to resist such forces unless they can be resisted by friction on the soil.

2.2 Exposure classification

Structural concrete elements are exposed to varying types of environmental conditions. The roof of a pumphouse is waterproofed with asphalt or roofing felt and, apart from a short period during construction, is never exposed to wet or damp conditions. The exposed legs of a water tower are

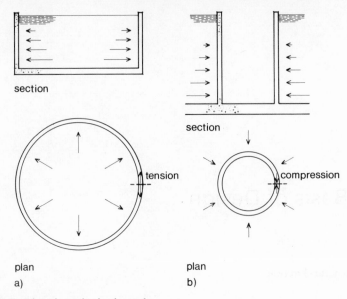

Figure 2.1 Direct forces in circular tanks
(a) Tensile forces
(b) Compressive forces

subjected to alternate wetting and drying from rainfall but do not have to contain liquid. The lower sections of the walls of a reservoir are always wet (except for brief periods during maintenance), but the upper sections may be alternately wet and dry as the water level varies. The underside of the roof of a closed reservoir is damp from condensation. These various conditions are illustrated in figure 2.4.

Experience has shown that, as the exposure conditions become more severe, precautions should be taken to ensure that moisture and air do not penetrate to the reinforcement and cause corrosion which, in turn, will

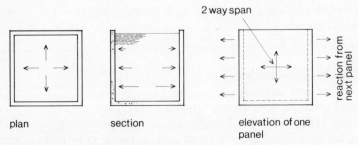

Figure 2.2 Direct forces of tension in wall panels of rectangular tanks

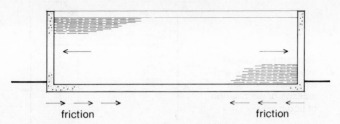

Figure 2.3 Tension in floor of a long tank with cantilever walls

cause the concrete surface to spall[15]. Adequate durability can be ensured by providing a dense well-compacted concrete mix (see section 2.5.2 and figure 2.5) with a concrete cover of 40 mm, but it is also necessary to control cracking in the concrete, and prevent percolation of liquid through the member.

It is convenient to classify exposure conditions for purposes of design, and to relate each class of exposure to a design crack width. In this book the classification adopted follows the designations in BS 5337[10] and is stated in table 2.1. The classification applies to each face of a concrete element, and an element may therefore have a different classification on

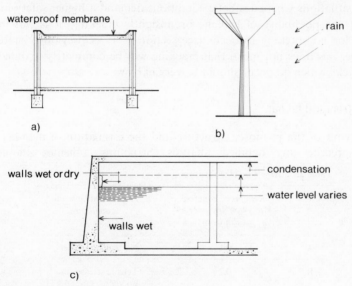

Figure 2.4 Exposure to environmental conditions
(a) Pumphouse roof
(b) Water tower
(c) Reservoir

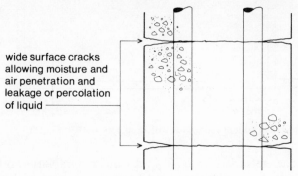

wide surface cracks
allowing moisture and
air penetration and
leakage or percolation
of liquid

Figure 2.5   Effect of cracks

each face. For relatively thin elements (up to 225 mm thick), it is preferable (and easier) to design each face for the most onerous condition on either face. Table 2.2 gives guidance on the appropriate classification for various conditions.

The application of this classification to a supply reservoir is shown in figure 2.6. It should be noted that, although the upper part of the inside of the wall is shown as Class A exposure, the conditions of stress normally allow design to Class B as indicated in chapter 3.

For situations where aesthetic conditions demand a higher safety margin against the possibility of leakage or unsightly cracking, it is possible to design for no cracking to occur (see section 3.7). As previously stated, in practice, this does not mean that cracking will be completely avoided, but any cracks which do occur should be very narrow.

2.3   Structural layout

The layout of the proposed structure and the estimation of member sizes must precede any detailed analysis. Structural schemes should be

Table 2.1   Exposure classification

Class	Reinforced concrete Maximum design crack width (mm)	Prestressed concrete condition
A	0.1	No tensile stresses
B	0.2	Tensile stresses but no visible cracking
C	0.3	Surface width of cracks not exceeding 0.2 mm

Table 2.2 Examples of exposure classification

Exposure class	Type of element
Class A	A liquid-retaining face which is exposed to alternate wetting and drying. A surface subject to condensation or freezing whilst wet.
Class B	A liquid-retaining face which is normally continuously wet.
Class C	Any normal non-liquid-retaining face not exposed to aggressive conditions, i.e. normal structural conditions.

considered from the viewpoints of strength, serviceability, ease of construction, and cost. These factors are to some extent mutually contradictory, and a satisfactory scheme is a compromise, simple in concept and detail. In liquid-retaining structures, it is particularly necessary to avoid sudden changes in section, because they cause concentration of stress and hence increase the possibility of cracking.

It is a good principle to carry the structural loads as directly as possible to the foundations, using the fewest structural members. It is preferable to design cantilever walls as tapering slabs rather than as counterfort walls with slabs and beams. The floor of a water tower or the roof of a reservoir can be designed as a flat slab. Underground tanks and swimming-pool tanks are generally simple structures with constant-thickness walls and floors.

It is essential for the designer to consider the method of construction and to specify on the drawings the position of all construction and movement joints. This is necessary as the detailed design of the structural elements will depend on the degree of restraint offered by adjacent sections of the structure to the section being placed. Important considerations are the provision of 'kickers' (or short sections of upstand concrete) against which formwork may be tightened, and the size of wall and floor panels to be cast in one operation (figure 2.10).

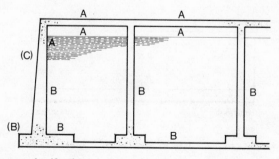

Figure 2.6 Exposure classification

2.4   Influence of construction methods

Designers should consider the sequence of construction when arranging the layout and details of a proposed structure. At the excavation stage, and particularly on water-logged sites, it is desirable that the soil profile to receive the foundation and floors should be easily cut by machine. Flat surfaces and long strips are easy to form but individual small excavations are expensive to form. The soil at foundation level exerts a restraining force on the structure which tends to cause cracking (figure 2.7). The frictional forces can be reduced by laying a sheet of 1000 g polythene or other suitable material on a 75 mm layer of 'blinding' concrete. For the frictional forces to be reduced, it is necessary for the blinding concrete to have a smooth and level surface finish. This can only be achieved by a properly screeded finish, and in turn this implies the use of a grade of concrete which can be so finished[16,17]. A convenient method is to specify the same grade of concrete for the blinding layer as is used for the structure. This enables a good finish to be obtained for the blinding layer, and also provides an opportunity to check the strength and consistency of the concrete at a non-critical stage of the job.

The foundations and floor slabs are constructed in sections which are of a convenient size and volume to enable construction to be finished in the time available. Sections terminate at a construction or movement joint (chapter 5). The construction sequence should be continuous as shown in figure 2.8 *a* and not as shown in figure 2.8 *b*. By adopting the first system, each section that is cast has one free end and is enabled to shrink on cooling without restraint (a day or two after casting). With the second method, considerable tensions are developed between the relatively rigid adjoining slabs.

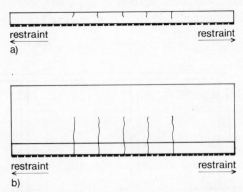

Figure 2.7   Cracking due to restraint by frictional forces at foundation level
(*a*) Floor slab
(*b*) Wall

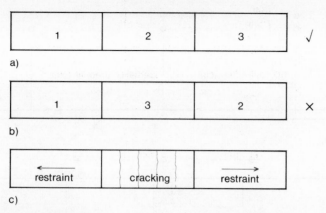

Figure 2.8 Construction sequence
(*a*) Preferred sequence
(*b*) Not recommended
(*c*) Effect of method (b) on third slab panel

The maximum spacing between movement joints depends on the amount of reinforcement provided, but generally about 7.5 m is the economical maximum distance between partial contraction joints, and 15.0 m between full contraction joints. Alternatively, temporary short gaps may be left out, to be filled in after the concrete has hardened. A further possibility is the use of induced contraction joints, where the concrete section is deliberately reduced in order to cause cracks to form at preferred positions. These possibilities are illustrated in figure 2.9. The casting sequence in the vertical direction is usually obvious. The foundations or floors are laid with a short section of wall to act as a key for the formwork (the kicker, figure 2.10). Walls may be concreted in one operation up to about 8 m height.

Reinforcement should be detailed to enable construction to proceed with a convenient length of bar projecting from the sections of concrete which are placed at each stage of construction. Bars should have a maximum spacing of 300 mm and a minimum spacing dependent on size, but not usually less than 100 mm to allow easy placing of the concrete. Distribution or shrinkage reinforcement should be placed in the outer layers nearest to the surface of the concrete. In this position it has maximum effect.

2.5 Materials and concrete mixes

2.5.1 *Reinforcement*[18]

Although the tensile stress in the reinforcement in liquid-retaining structures is relatively low, it is usual to specify high-strength steel with a

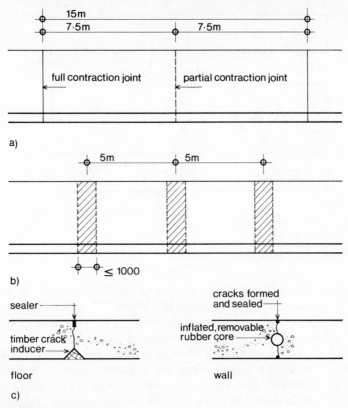

Figure 2.9 Joints
  (a) Typical layout of joints in a wall
  (b) Typical layout of temporary gaps in construction
  (c) Induced joints

ribbed or deformed surface. The difference in cost between high-strength ribbed steel and plain-surface mild steel is only about 3% (UK). This small extra cost is more than saved by the extra strength available and increased bond performance. Similar arguments affect the use of welded fabric reinforcement, where fixing costs are very much reduced and time saved. Traditionally, fabric has been used only in ground slabs but, where the quantity is sufficient, can now be obtained in sizes and types that allow it to be used in walls, floors and roofs.

The specified characteristic strengths of reinforcement available in the UK are given in table 2.3. The specified characteristic strength is a statistical measure of the yield or proof stress of a type of reinforcement. For bars supplied in accordance with British Standards, the proportion of

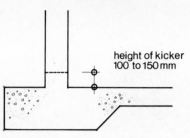

Figure 2.10   Joint between floor and wall

bars which fall below the characteristic strength level is defined as 5%
(figure 2.11)[19,20,21]. A material partial safety factor $\gamma_m = 1.15$ is applied to
the specified characteristic strength to obtain the ultimate design strength.
In BS 5337, an arbitrary upper limit of 425 N/mm² is also applied to the
characteristic strength used in design to avoid development of high service
stresses.

   Reinforcement embedded in concrete is protected from corrosion by the
alkalinity in the cement. If particularly aggressive conditions prevail, it may
be desirable to use galvanized reinforcement, especially when small
thicknesses of concrete are used.

2.5.2   Concrete
The detailed specification and design of concrete mixes is outside the scope
of this book, but the essential features of a design are given below.
Guidance on mix design may be found from the references[16,17].

Cements. Normal Portland cement is generally used for liquid-retaining
structures. It is not desirable to use rapid-hardening cement because of its
greater evolution of heat which tends to increase shrinkage cracking.
However, its use may be considered in cold weather. When there are
sulphates in the ground water, or other chemical contaminants, the use of
sulphate-resisting cement or super-sulphated cement may be essential.

Table 2.3   Types and strengths of reinforcement (UK)

BS	Type	Characteristic yield or proof stress f_y (N/mm²)
4449	Plain mild steel	250
4449	Hot-rolled high-yield ribbed bars	460/425*
4461	Cold-worked high-yield ribbed bars	460/425*
4483	Welded fabric	485

* Sizes 8–16 = 460. Sizes 20–40 = 425

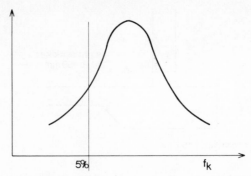

Figure 2.11   Graphical definition of characteristic strength

Aggregates. The maximum size of aggregate must be chosen in relation to the thickness of the structural member. 20 mm maximum size is always specified up to member thickness of about 300–400 mm, and may be used above this limit. Size 40 may be specified in very thick members. The use of a large maximum size of aggregate has the effect of reducing the cement content in the mix for a given workability, and hence reduces the amount of shrinkage cracking.

It is important to choose aggregates that have low drying shrinkage and low absorption. Most quartz aggregates are satisfactory in these respects but, where limestone aggregate is proposed, some check on the porosity is desirable. Certain aggregates obtained from igneous rocks exhibit high shrinkage properties and are quite unsuitable for use in liquid-retaining structures[22].

Local suppliers can often provide evidence of previous use which will satisfy the specifier, but some care is necessary in using material from a new quarry, and tests of the aggregate properties are recommended.

Admixtures. Admixtures containing appreciable amounts of calcium chloride are not desirable as there is a risk of corrosion of the reinforcement. Other admixtures which improve workability or frost resistance may be used on their merits[44].

Concrete mix design. The concrete must be designed to provide a mix which is capable of being fully compacted by the means available. Any areas of concrete which have not been properly compacted are likely to leak. The use of poker-type internal vibrators is recommended.

The cement content in kg/m³ of finished concrete must be judged in

Table 2.4 Cement content in concrete
Minimum cement content in kg/m^3 (to ensure durability)

Exposure class	Reinforced concrete	Prestressed concrete
A	360 kg/m^3	360 kg/m^3
B	290 kg/m^3	300 kg/m^3

(20 mm aggregate).

relation to a minimum value to ensure durability, and a maximum value to avoid a high temperature rise in the freshly placed concrete. Precise values depend on cement quality and strength, but UK practice is shown in table 2.4.

Maximum cement content. For reinforced concrete, the cement content should not exceed 400 kg/m^3; for the higher grades of prestressed concrete, higher cement contents up to a maximum of 550 kg/m^3 may be used.

2.6 Loading

Liquid-retaining structures are subject to loading by pressure from the retained liquid. Typical values of weights are listed in table 2.5.

The designer must consider whether sections of the complete reservoir may be empty when other sections are full, and design each structural element for the maximum bending moments and forces that can occur. Several loading cases may need to be considered. Internal partition walls should be designed for liquid loading on one side only.

External reservoir walls are often required to support soil fill. The loading conditions to be considered are illustrated in figure 2.12. When the reservoir is empty, full allowance must be made for the active soil pressure, and any surcharge pressures from vehicles. It is important to note that when

Table 2.5 Density of retained liquids

Liquid	Weight (kN/m^3)
Water	10.0
Raw sewage	11.0
Digested sludge aerobic	10.4
Digested sludge anaerobic	11.3
Sludge from vacuum filters	12.0

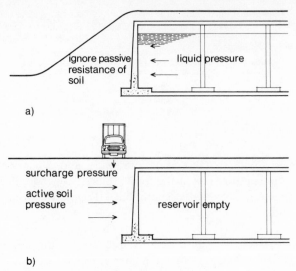

Figure 2.12   Design loadings for external walls with soil fill
(*a*) Reservoir full
(*b*) Reservoir empty

designing for the condition with the reservoir full, no relief should be allowed from passive pressure of the soil fill. This is because of the differing moduli of elasticity of soil and concrete which prevent the passive resistance of the soil being developed before the concrete is fully loaded by the pressure from the contained liquid[23].

2.6.1   Partial safety factors for loads

When designing a structural element for the ultimate limit state, it is necessary to use partial safety factors (in conjunction with the characteristic applied loading) to provide the necessary margin against failure. The factors take account of the likely variability of the loading and the consequences of failure.

For dead loads a factor $\gamma_f = 1.4$ may be used, and for imposed loads $\gamma_f = 1.6$. Although the loading due to water is known precisely, it is recommended in BS 5337 that $\gamma_f = 1.6$ is used for loads due to retained liquids. In practice, this is not a restriction, as the crack width calculation at service loads controls the design; and it is not likely that a satisfactory structure would result from the use of a lower load factor, as the service stress would be very high.

Given the present state of knowledge, $\gamma_f = 1.6$ must also be used with active pressures due to soil.

2.7 Foundations

It is desirable that a liquid-retaining structure is founded on good uniform soil, so that differential settlements are avoided (chapter 1). However, this desirable situation is not always obtainable. Variations in soil conditions must be considered, and the degree of differential settlement estimated[24]. Joints may be used to allow a limited degree of articulation but, on sites with particularly non-uniform soil, it may be necessary to consider dividing the structure into completely separate sections. Alternatively, cut-and-fill techniques may be used to provide a uniform platform of material on which to found the structure.

Soils which contain bands of peat or other very soft strata may not allow normal support without very large settlements, and piled foundations are required[24].

The design of structures in areas of mining activity requires the provision of extra joints, or the division of the whole structure into smaller units. Pre-stressed tendons may be added to a normal reinforced concrete design to provide increased resistance to cracking when movement takes place[13].

The use of cantilever walls depends on passive resistance to the applied loads, resistance to sliding being provided by the foundation soil. If the soil is inundated by ground water, it may not be possible to develop the necessary soil pressure under the footing. In these circumstances, a cantilever design is not appropriate, and the overturning forces should be resisted by a system of beams balanced by the opposite wall, or by designing the wall to span horizontally if that is possible.

Walls which are designed as propped cantilevers, using the roof structure as a prop, are often considered to have no rotation at the footing (figure 2.13). The strain in a cohesive soil may allow some rotation and a

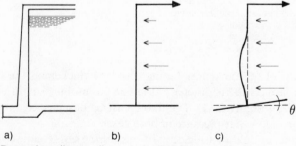

a)   b)   c)

Figure 2.13 Propped cantilever walls on a cohesive soil
  (a) Structure
  (b) Basic structural assumptions
  (c) Rotation due to soil movement

redistribution of forces and moments. An example of this calculation is given in chapter 6.

2.8   Flotation

An empty tank constructed in water-bearing soil will tend to move upwards in the ground, or float. This tendency must be counteracted by ensuring that the weight of the empty tank structure is greater than the uplift equal to the weight of the ground water displaced by the tank. The safety margin required is a matter for the judgement of the designer. Dependent on the certainty with which the ground water level is known, the factor of safety may vary between 1.05 and 1.25.

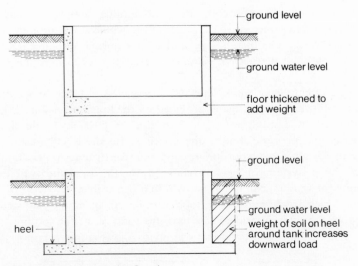

Figure 2.14   Methods of preventing flotation
  (*a*) Additional dead weight
  (*b*) Provision of a heel

The weight of the tank may be increased by thickening the floor or by providing a heel on the perimeter of the floor to mobilize extra weight from the external soil (figure 2.14). The floor must be designed against the uplift due to the ground water pressure in both cases.

The designer should consider conditions during construction, in addition to the final condition, and specify a construction sequence to ensure that the structure is stable at each phase of construction.

An example of the calculation is now given.

Example 2.1.

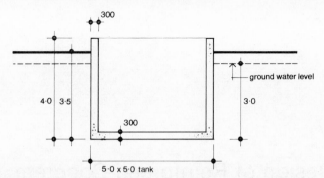

A tank of overall size 5.0 m × 5.0 m × 4.0 m deep is to be constructed with the underside of the floor at a level 3.5 m below ground level. The walls and floor are 300 mm thick.

Check stability against flotation if the ground water level is 0.5 m below the soil surface, and the required factor of safety against flotation is 1.15.

Weight of empty tank:

Floor	$24 \times 0.3 \times 5.0 \times 5.0$	$= 180$
Walls	$24 \times 0.3 \times 4.7 \times 4.0 \times 3.7$	$= 500$
	Total weight of empty tank	680 kN

Uplift due to ground water
$$= 10 \times 5.0 \times 5.0 \times 3.0 \qquad = 750 \, kN$$
Required dead weight $= $ (factor of safety) × (uplift)
$$= 1.15 \times 750 \qquad\qquad = 863 \, kN$$
∴ extra weight required $= 853\text{--}680$
Try two possible solutions.
(*a*) Increase floor thickness to 375 mm.
Extra weight provided
$$= 24 \times 0.075 \times 5.0 \times 5.0 \qquad = 225 \, kN$$

This is satisfactory.

(*b*) Provide heel on outside of floor.

Provide a heel 0.5 m wide around outside walls.

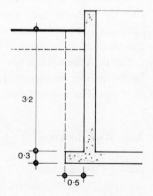

Weight of soil	$= 18 \, kN/m^3$
Weight of soil submerged in ground water	
$= 18\text{--}10$	$= 8 \, kN/m^3$
Weight of soil carried by heel if tank attempts to lift	
$= 8 \times 0.5 \times 3.2$	$= 12.8 \, kN/m \text{ run}$
Perimeter of tank $= 4 \times 5.0$	$= 20 \, m$
Total weight of soil $= 12.8 \times 20$	$= 256 \, kN$

This is satisfactory.

Note: For simplicity, it is assumed that all the soil is submerged in water. This is conservative, as the upper 0.5 m is above ground water level.

3 Design of Reinforced Concrete

3.1 General[26,9]

The basic design philosophy of liquid-retaining structures is discussed in chapter 2. In this chapter, detailed design methods are described to ensure compliance with the basic requirements of strength and serviceability.

In contrast with normal structural design, where strength is the basic consideration, for liquid-retaining structures it is found that serviceability considerations control the design. The procedure is therefore:

(a) Decide on concrete member sizes.
(b) Calculate the reinforcement required to limit the design crack widths to the required value.
(c) Check strength.

The calculation of crack widths in a member subjected to flexural loading can be carried out once the overall thickness and the quantity of reinforcement have been determined, but it is not possible to make a direct calculation. It is therefore convenient to use design tables or charts. The tables in Appendix A are arranged to enable the whole structural design to be carried out in one operation, including the checking of crack control and strength. The use and derivation of the tables is described in the Appendix.

3.2 Wall thickness

3.2.1 Considerations

All liquid-retaining structures include wall elements to contain the liquid, and it is necessary to commence the design by estimating the overall wall thickness in relation to the height.

The overall thickness of a wall should be no greater than necessary, as extra thickness will cause higher thermal forces when the concrete is hardening. The principal factors which govern the wall thickness are:

(a) ease of construction
(b) strength in shear
(c) avoidance of excessive deflections
(d) adequate strength
(e) avoidance of excessive crack widths

The first estimate of minimum section thickness is conveniently made by considering (a), (b) and (c). The results of this preliminary estimation are given in table 3.1. Factors (d) and (e) are considered in sections 3.3 and 3.4.

Table 3.1 Approximate minimum thickness h (mm) of R. C. Cantilever wall subjected to water pressure.

$\dfrac{100A_s}{bd}$	0.25	0.5	1.0
Height of wall (m)			
8	1550	1100	875
6	900	650	525
4	450	325	275
2	200	200	200

3.2.2 Ease of construction

If a wall is too thin in relation to its height, it will be difficult for the concrete to be placed in position and properly compacted. As this is a prime requirement for liquid-retaining structures, it is essential to consider the method of construction when preparing the design. It is usual to cast walls up to about 7 metres high in one operation and, to enable this to be successfully carried out, the minimum thickness of a wall over 2 metres high should be not less than 250–300 mm. Walls less than 2 metres high may have a minimum thickness of 200 mm. A wall thickness less than 200 mm is not normally possible, as the necessary four layers of reinforcement cannot be accommodated with the appropriate concrete cover on each face of the wall. The wall may taper in thickness with height in order to save materials. Setting out is facilitated if the taper is uniform over the whole height of the wall (figure 3.1).

3.2.3 Strength in shear

It is inconvenient to use shear reinforcement in slabs because it is difficult to fix, impedes placing of the concrete, and is inefficient in the use of steel. The

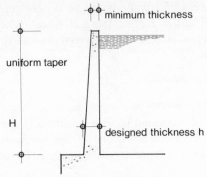

Figure 3.1   Typical section through a wall

wall thickness should therefore be at least sufficient to allow the ultimate shear forces to be resisted by the concrete in combination with the longitudinal reinforcement. It is, in theory, also necessary to ensure that any diagonal cracks due to shear at service loads are within the allowable limits, but in practice, other requirements will ensure that no check need be made. Suitable values of the ultimate shear stress on the concrete in slabs are given in table 3.2. As this stress varies, not only with concrete grade but also with the reinforcement ratio in the section, some estimate of the reinforcement percentage must be made at the outset. For cantilever walls, a value of 0.5% is a reasonable starting point. Design tables may be constructed from the values in table 3.2 as follows:

Table 3.2   Allowable ultimate shear stress in slabs (N/mm²)

$\dfrac{100A_s}{bd}$	Concrete grade	
	25	30
0.25	0.35	0.35
0.50	0.50	0.55
1.00	0.60	0.70
2.00	0.85	0.90

Note: These values follow those in table 5 of CP 110, but are slightly conservative, as the ratios for slabs from table 14 of the Code have not been included.

$$\text{The steel ratio} = \frac{100A_s}{bd}$$

where A_s is the area of the longitudinal tension steel that is fully anchored.

Assume a free cantilever wall of uniform tapered section subjected to water pressure (figure 3.2)

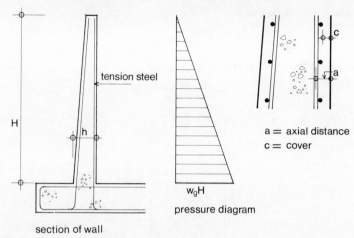

a = axial distance
c = cover

pressure diagram

section of wall

Figure 3.2 Cantilever wall subjected to water pressure

H = height of liquid (m)
w_g = density of liquid (kN/m³)
h = maximum overall thickness of section (mm)
d = maximum effective depth of section (mm)
a = depth of centre of tension steel from face of concrete (mm)
The overall thickness $h = d + a$
γ_f = partial safety factor for loads
The maximum ultimate shear force $V_u = K_r w_g \gamma_f H^2$ where $K_r = 0.5$ for a cantilever wall

The shear stress on the section $v = \dfrac{V_u}{bd}$

The distance from the face of the concrete to the centre of the tension steel a varies according to bar size and cover. Assuming that the concrete cover is 40 mm and the bar size is 16 mm, the value of a is equal to $40 + 1.5 \times 16$ or about 65 mm. (Distribution reinforcement should be in the outer layer where it is more effective.) The required section thickness h may be calculated from given values of applied shear force and permissible stress to ensure that no shear reinforcement is required. An example of the calculation follows.

Example 3.1

Calculation of wall thickness for shear strength
Consider a cantilever wall of height H subject to water pressure.

Height $H = 6.0$ m.
Density of water $= w_g = 10$ kN/m³.
Partial safety factor $\gamma_f = 1.6$.

Assume tension reinforcement ratio

$$\frac{100A_s}{bd} = 0.5\%$$

Maximum applied ultimate shear force at base level

$$= V_u = 0.5 \times 10 \times 1.6 \times 6.0^2$$
$$= 288\,\text{kN/m}$$

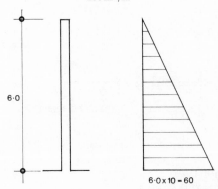

6·0

6·0 x 10 = 60

For grade 25 concrete, allowable shear stress $= 0.5\,\text{N/mm}^2$ (table 3.2). The minimum value of d is therefore

$$d = \frac{288 \times 10^3}{0.5 \times 10^3} = 576\,\text{mm}$$

and the overall wall thickness

$$h = d + a = 576 + 60 = 636\,\text{mm}$$

From shear considerations, the wall should have a minimum thickness of (say) 650 mm.

3.2.4   Deflection

The lateral deflection of a cantilever wall which is proportioned according to the rules suggested in this chapter is likely to be no more than about 30 mm. A wall which is restrained by connection to a roof slab or lateral walls will clearly deflect even less. Deflection of this magnitude will have no effect on the containment of liquid and, unless there is a roof slab supported by the wall with a sliding joint, there is no need to consider the amount of deflection. If pipes or other apparatus pass through a wall which may itself move slightly under load, the pipes must be arranged to be sufficiently flexible to allow for this movement.

Concrete codes allow members to have stiffness defined in terms of span/effective depth ratios as an alternative to calculating deflections. These values apply equally to normal and liquid-retaining structures[6]. Typical values are given in table 3.3 (from CP 110). The values are based on limiting the deflections to span/250, assuming that the member is constant in depth and that the loading is uniform. In the case of a vertical cantilever wall subjected to liquid pressure, the loading will be of triangular distribution and the wall section may be tapered. If the values in table 3.3 are used as the basis for calculating the effective depth of the member, a slightly conservative design will result. Allowance may be made for the effect of the triangular load distribution by increasing the basic allowable ratio for a cantilever from 7 to 8.75. (This is based on a comparison of deflection coefficients.)

Table 3.3 Span/effective depth ratios for slabs up to 10 m span. (The effect of compression reinforcement has not been taken into account.)
(a) Basic Ratios

Condition	Ratio
Cantilever	7
Simply supported	20
Continuous	26

(b) Modification factors for tension reinforcement (at service stress $f_s = 200\,\text{N/mm}^2$)

$\dfrac{100A_s}{bd}$	0.25	0.50	0.75	1.0	1.15
Factor	2.0	1.46	1.26	1.15	1.02

The discussion above deals with the deflection of cantilevers assuming a fixed base, but further lateral deflection may be caused by rotation of the base due to consolidation of the soil. This factor is of importance for high walls and relatively compressible ground. An estimation of the lateral deflection at the top of a wall due to base rotation may be made by considering the vertical displacements of the extremities of the foundation with the reservoir full, assuming that the wall and base are rigid, and subjected to rotation calculated from the differential soil consolidation at front and rear of the footing (figure 3.3).

$$\text{Rotation } \phi = \frac{a_1 - a_2}{B} \quad \text{and} \quad a_r = \phi H$$

The value must be added to the deflection due to the flexure of the wall calculated by

$$a_w = \frac{w_g H^5}{30EI}$$

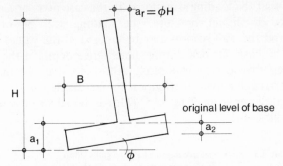

Figure 3.3   Rotation of cantilever wall due to soil consolidation

Finally, the total deflection at the top of the wall

$$a = a_w + a_r$$

is compared with $H/250$ or any other requirement.

With a propped cantilever wall, deflection will not be critical, but the rotation of the base will alter the relation between the negative and positive moments in the wall. The moments may be calculated most easily using a computer program[25].

3.3   Design considerations

If a reinforced concrete slab is laterally loaded, the concrete on the side of the tension reinforcement will extend and, dependent on the magnitude of loading (other factors being equal), it will eventually crack as the load is increased. Further increases in load widen the cracks that have formed and increase the stress in the reinforcement (figure 3.4). If the load is kept constant and the amount of reinforcement is increased, the stresses in the steel will be reduced, and the crack widths will be narrower.

Using limit state design, the applied load is fixed by the structural arrangement, and the designer has to choose values of slab thickness, reinforcement quantity, and reinforcement service stress to ensure that the crack widths under service loads are within the appropriate values given by the grade of exposure (chapter 2), and that the ultimate load factor is satisfactory.

Although a crack width calculation may show that reinforcement service stresses as high as $250\,N/mm^2$ are possible with certain combinations of slab thickness and reinforcement, it is not advisable to choose these arrangements. It is suggested that an arbitrary upper limit of $200\,N/mm^2$ is placed on the value of reinforcement service stress. The values given in

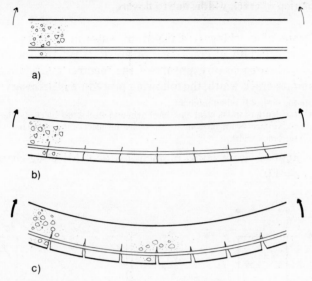

Figure 3.4 Flexural cracking
  (*a*) Concrete uncracked with low steel stress
  (*b*) Fine cracks and increased steel stress
  (*c*) Wide cracks and high steel stress

Appendix A take account of this restriction, although some values slightly over a service stress of $200 \, \text{N/mm}^2$ are given to allow interpolation in the tables.

There is no single design that will simultaneously exactly meet all the required criteria, and a number of different solutions are possible, even for a given value of design crack width. The tables in Appendix A have been constructed to allow the designer to choose directly a section thickness and arrangement of reinforcement at a stated service stress.

The derivation of the tables is given in Appendix A. The tables assume various values of section thickness, cover, and reinforcement size and spacing. The value of reinforcement stress is then calculated for a stated value of crack width. The tables have been prepared for concrete grade 25 and steel grade 425, but may be used for concrete grade 30 with little loss of accuracy.

The detailed methods of calculation for limit state design are considered in sections 3.4 and 3.5. Where direct tensile forces are present in addition to flexural forces, it is not possible at present to calculate crack widths, and an alternative design method must be used by limiting the stresses in the tensile reinforcement to a specified value taking account of the exposure conditions. This design method is considered in section 3.6.

3.4　Calculation of crack widths due to flexure

The limit state of cracking is satisfied by ensuring that the maximum calculated surface width of cracks is not greater than the specified value depending on the degree of exposure of the member (see chapter 2). To check the surface crack width, the following procedure is necessary:

(a) Calculate the service bending moment.
(b) Calculate the depth of the neutral axis, lever arm and steel stress by elastic theory.
(c) Calculate the surface strain allowing for the stiffening effect of the concrete.
(d) Calculate the crack width.

The maximum service bending moment is calculated using characteristic loads with $\gamma_f = 1.0$.

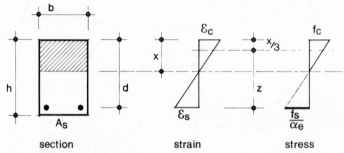

| section | strain | stress |

Figure 3.5　Assumed stress and strain diagrams—cracked section—elastic design

The depth of the neutral axis x is calculated (see section 3.6.1) using the usual formula for modular ratio design (figure 3.5):

$$\frac{x}{d} = \alpha_e\rho\left(\sqrt{1 + \frac{2}{\alpha_e\rho}} - 1\right)$$

α_e = modular ratio = $\dfrac{E_s}{E_c}$ (*Note:* E_c should be taken as half the instantaneous value.)

$\rho = \dfrac{A_s}{bd}$

Typical values of the short-term modulus of elasticity for concrete are given in table 3.4.

A similar but more complex formula may be used when compression reinforcement is present. From x the lever arm z is found from

$$z = d - \frac{x}{3}$$

Table 3.4 Values of short-term modulus of elasticity of concrete

Concrete grade	Modulus of elasticity
25	26
30	28

The tensile steel and concrete compressive stresses are then

$$f_s = \frac{M_s}{zA_s}$$

$$f_{cb} = \frac{2M_s}{zbx}$$

For the crack width formula to be valid, the compressive stress in the concrete and the tensile stress in the steel under service conditions must be less than limiting values as follows:

$$\text{Concrete:} \quad f_{cb} \not> 0.45\, f_{cu}$$
$$\text{Steel:} \quad f_s \not> 0.8\, f_y$$

If these criteria are met, the formulae for crack width calculation may be used.

The average strain at the surface is calculated from the formula[10] (figure 3.6):

$$\varepsilon_m = \varepsilon_1 - \frac{0.7b_t h(a'-x)}{A_s(h-x)f_s} \times 10^{-3}$$

The maximum width of a surface crack in a slab is along a line mid-way between two adjacent bars[6] (figure 3.7).

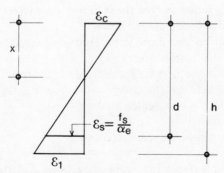

Figure 3.6 Crack calculation—strain diagram

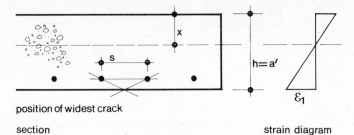

position of widest crack

section   strain diagram

Figure 3.7   Slab section indicating surface crack

For a slab, $(a - x) = (h - x)$ and the formula reduces to

$$\varepsilon_m = \varepsilon_1 - \frac{0.7b_t h}{A_s f_s} \times 10^{-3}$$

The value of ε_1 represents the calculated average elastic tensile strain in the concrete (using 'no tension' theory), and the second term makes allowance for the stiffening effect of the concrete between actual cracks.

$$\varepsilon_1 = \frac{(h - x)}{(d - x)} \times \frac{f_s}{E_s}$$

The design surface crack width is obtained from the formula

$$\text{design surface crack width } w = \frac{4.5 a_{cr} \varepsilon_m}{1 + 2.5\left(\dfrac{a_{cr} - c_{min}}{h - x}\right)}$$

which is stated in BS 5337.

It should be noted that this formula is similar in form to the equivalent formula in CP 110, but the coefficients are different. The coefficients in the formula from BS 5337 have been chosen to allow for a greater statistical certainty that the calculated crack width will not be exceeded. A negative calculated value of w indicates that the section is uncracked.

The symbols used are summarized below:

ε_1 is the strain at the level considered, ignoring the stiffening effect of the concrete in the tension zone.

b_t is the width of the section at the centroid of the tension steel.

a' is the distance between the compression face and the point at which the crack width is being calculated.

A_s is the area of steel.

f_s is the service stress in the reinforcement.

a_{cr} is the distance between the point considered and the surface of the nearest longitudinal bar.

ε_m is the average strain at the level at which cracking is being considered, allowing for the stiffening effect of the concrete in the tension zone.

c_{min} is the minimum cover over the tension steel.

h is the overall depth of the member.

x is the depth of the neutral axis.

E_s is Young's modulus for steel.

E_c is Young's modulus for concrete.

An example of a crack width calculation is given below.

Example 3.2
To calculate the design crack width in a flexural member.
A wall of 300 mm overall thickness is reinforced with Y16 bars at 200 mm centres.

Cover = 50 mm   Concrete grade 25

Calculate the design crack width for an applied service moment of 44 kN m/m.

Reinforcement area = 1005 mm^2/m
$d = 300 - 50 - 8 = 242$ mm

$$\rho = \frac{A_s}{bd}$$

$$= \frac{1005}{1000 \times 242}$$

$$= 0.00415$$

Modular ratio:

$$E_s = 200 \, kN/mm^2$$

$$E_c = \tfrac{1}{2} \times 26 = 13 \, kN/mm^2$$

$$\alpha_e = E_s/E_c = 15$$

$$\alpha_e \rho = 0.0623$$

Depth of neutral axis is given by

$$\frac{x}{d} = 0.0623 \left(\sqrt{1 + \frac{2}{0.0623}} - 1 \right)$$

$$= 0.296 \quad \text{and} \quad x = 71.6 \, mm$$

Lever arm $= z = d - \dfrac{x}{3}$

$$= 242 - 24 = 218 \, mm$$

Steel tensile stress $f_s = M_s/(zA_s)$

$$f_s = 44 \times 10^6/(218 \times 1005) = 201 \, N/mm^2$$

Concrete compressive stress

$$f_{cb} = \frac{2M_s}{zbx} = \frac{2 \times 44 \times 10^6}{218 \times 10^3 \times 71.6} = 5.64 \, N/mm^2$$

Check stress levels:

$$0.45 f_{cu} = 0.45 \times 25 \qquad = 11.25$$
$$f_{cb} = 5.64 < 11.25 \qquad \text{O.K.}$$
$$0.8 f_y = 0.8 \times 425 \qquad = 340$$
$$f_s = 201 < 340 \qquad \text{O.K.}$$

$$\varepsilon_1 = \frac{h-x}{d-x} \times \frac{f_s}{E_s}$$

$$= \frac{300-71.6}{242-71.6} \times \frac{201}{200 \times 10^3} = 1.347 \times 10^{-3}$$

$$\text{Correction factor} = \frac{0.7 \times b_t h}{A_s f_s} \times 10^{-3}$$

$$= \frac{0.7 \times 10^3 \times 300}{1005 \times 201} \times 10^{-3}$$

$$= 1.040 \times 10^{-3}$$

Average surface strain

$$\varepsilon_n = (1.347 - 1.040) \times 10^{-3} = 0.307 \times 10^{-3}$$

Design surface crack width

$$w = \frac{4.5 a_{cr} \varepsilon_n}{1 + 2.5 \left(\dfrac{a_{cr} - c_{min}}{h-x} \right)}$$

$$a_{cr} = \sqrt{\left(\frac{s}{2}\right)^2 + \left(c + \frac{\phi}{2}\right)^2} - \frac{\phi}{2}$$

$\phi = 16$
$c = 50 = c_{min}$
$s = 200$
$a_{cr} = 107.6\,\text{mm}$

hence

$$w = \frac{4.5 \times 107.6 \times 0.307 \times 10^{-3}}{1 + 2.5 \left(\dfrac{107.6 - 50}{300 - 71.6} \right)} = 0.091\,\text{mm}$$

3.5 Strength calculations

The analysis of the ultimate flexural strength of a section is made using formulae applicable to the design of normal structures, but the design characteristic strength of the reinforcement is limited to $425\,\text{N/mm}^2$. This is an arbitrary upper limit, but is not restrictive in practice due to the need to control cracking. The partial safety factor for loads due to liquid pressure is taken as $\gamma_f = 1.6$.

Although the liquid load is known almost precisely, the possibility of failure is usually of consequence, and therefore $\gamma_f = 1.6$ is appropriate, and is not in practice restrictive.

The formulae for the calculation of the ultimate limit state condition are obtained from a consideration of the forces of equilibrium and the shape of the concrete stress block at failure.

The partial safety factor for concrete is taken as $\gamma_c = 1.5$ and for steel $\gamma_s = 1.15$. After allowing for the partial safety factor for concrete, for the UK practice of testing concrete strength using cubes, and for the equivalent rectangular stress block, a value of $0.45 f_{cu}$ is used for the width of the stress block.

Using the rectangular stress block as illustrated in figure 3.8, the following equations may be derived:

$$\text{Lever arm} = z_1 = 1 - 0.45x_1 \tag{1}$$

Force of tension = force of compression

$$\therefore \quad \frac{A_s f_y}{1.15} = 0.45 f_{cu} b \times 0.9x \tag{2}$$

Moment of resistance based on steel

$$M = \frac{f_y}{1.15} A_s z_1 d \tag{3}$$

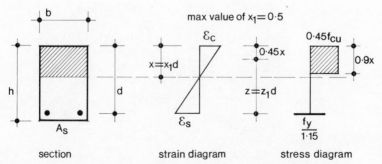

Figure 3.8 Assumed stress and strain diagrams—ultimate flexural limit state design

With the maximum permissible value of $x = d/2$, the moment of resistance based on the concrete section is

$$M_u = 0.157 f_{cu} b d^2 \qquad (4)$$

M_u represents the maximum ultimate moment which can be applied to the section without using compression reinforcement. The actual applied ultimate moment M will generally be less than M_u.

By re-arranging equations (1)–(4) and solving for the depth of the neutral axis

$$x_1 = \left(1 - \sqrt{1 - 0.7\,\frac{M}{M_u}}\right)\bigg/ 0.9 \qquad (5)$$

This value may be substituted in equations (1) and (3) to calculate the required area of reinforcement.

After the arrangement of reinforcement has been decided, the ultimate shear stress is checked using

$$v = \frac{V_u}{bd}$$

The actual ultimate shear stress v should be less than the allowable ultimate shear stress given in table 3.2.

An example of a typical strength calculation follows.

Example 3.3
Strength calculation—Limit state design
Calculate the wall thickness and reinforcement required to provide the necessary load factor for a wall subjected to water pressure over a height of 2.9 m ($\gamma_f = 1.6$).

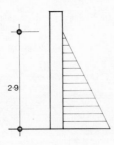

Applied ultimate moment

$$M_u = \tfrac{1}{6}(10 \times 1.6) \times 2.9^3$$
$$= 65.0\,\text{kN m/m}$$

Applied ultimate shear force

$$V_u = \tfrac{1}{2}(10 \times 1.6) \times 2.9^2$$
$$= 67.3\,\text{kN/m}$$

From table 3.2, for an assumed steel ratio of 0.25%, the allowable ultimate shear stress = 0.35.

$$\text{Required effective depth } d = \frac{0.35 \times 10^3}{67.3}$$

$$= 193\,\text{mm}$$

$$\text{Use wall thickness } h = 193 + \text{cover} + \frac{\phi}{2}$$

$$\text{Say } \; h = 300\,\text{mm}$$

Overall wall thickness $h = 300$.
Assume bar size 16 mm.
Assume cover to distribution steel = 40 mm.
Assume distribution steel size 12.

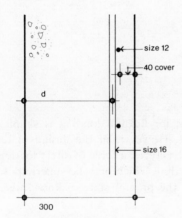

$$d = 300 - 40 - 12 - 8$$
$$= 240\,\text{mm}$$

$$M_u = 0.157 f_{cu} b d^2$$
$$= 0.157 \times 2.5 \times 10^3 \times 240^2 \times 10^{-6}$$
$$= 226\,\text{kN m.}$$

Design ultimate moment = 65 kN m

$$\text{Lever arm} = x = \left[\left(1 - \sqrt{1 - 0.7\,\frac{M}{M_c}} \right) \Big/ 0.9 \right] d$$

$$= 0.12 \times 240 = 28.8$$

$$z = d - \frac{x}{2} = 225$$

Area of tensile steel/metre width

$$A_s = \frac{1.15M}{f_y z}$$
$$= \frac{1.15 \times 65 \times 10^6}{425 \times 225} = \underline{782\,\text{mm}^2}$$

This is the minimum area of steel to give the specified load factor for the ultimate limit state. A suitable arrangement of reinforcement is Y16 at 250 (804 from bar table) and this must be checked for crack width at working load.

Shear:

$$\text{Steel ratio } \rho = \frac{A_s}{bd} = \frac{804}{10^3 \times 240} = 0.34\%$$

Allowable shear stress on concrete

$$= v_c = 0.40\,\text{N/mm}^2$$
$$\text{Actual shear stress} = \frac{V_u}{bd}$$
$$= \frac{67.3 \times 10^3}{10^3 \times 240}$$
$$= 0.28\,\text{N/mm}^2$$

Satisfactory

3.6   Limiting stress design

The limit state method of design for flexural loading is simple to apply when design tables are available. However, in the design of rectangular tanks and certain other types of structures, a force of direct tension exists in the member due to structural loading which may be superimposed on the flexural action (figure 3.9). With the present state of knowledge, it is not

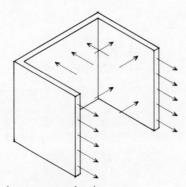

Figure 3.9   Tensile forces due to structural action

possible to calculate crack widths with the same degree of certainty as for purely flexural cracking. Consequently full limit state design procedures cannot be applied.

In these circumstances, crack widths may be controlled by limiting the stress in the tensile reinforcement under service conditions to the values given in table 3.5. This method of satisfying the limit state crack control requirements is sometimes known as the 'deemed to satisfy' method of design, in that use of this procedure is deemed to satisfy the crack width limitation requirement.

The calculations are prepared using the elastic modular ratio method of design. For an accurate analysis, the depth of the neutral axis must be calculated for the applied direct tensile force and bending moment, but for many structures it will be safe, and sufficiently accurate, to consider separately the effects for each applied force. The calculated quantities of reinforcement are then added together. The necessary formulae are given in sections 3.6.1 and 3.6.2, and an example of a design is given below (Example 3.4).

Table 3.5 Allowable steel stresses in direct or flexural tension for serviceability limit states

BS 5337[10]

Class of exposure	Allowable stress (N/mm^2)	
	Plain bars	Deformed bars
A	85	110
B	115	130

ACI 350R-77[11]

Exposure	Bar sizes* (mm)	Allowable stress* (N/mm^2)
A*	10–16	151
	20–25	124
	28–35	117
B*	10–16	186
	20–25	151
	28–35	144
Direct tension	all sizes	96

Note: The exposure classifications in ACI code are not identical with BS 5337 but are broadly similar in intention. For purposes of comparison, U.S. bar sizes and stresses have been translated into metric values and rounded. The ACI code also limits the spacing of the reinforcement.

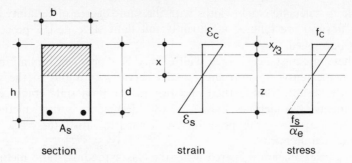

Figure 3.10   Assumed stress and strain diagrams—cracked section—elastic design

3.6.1   *Flexural reinforcement*

Figure 3.10 illustrates the assumptions made using the elastic modular ratio method of design with a cracked section and no tensile force in the concrete.

The depth of the neutral axis depends on the steel ratio ρ and may be obtained by equating the force of tension in the steel with the force of compression in the concrete:

$$f_s A_s = 0.5 f_{cb} b x \tag{1}$$

From the strain diagram:

$$\frac{f_s/E_s}{d-x} = \frac{f_{cb}/E_c}{x} \tag{2}$$

From (1)

$$f_s/f_{cb} = \frac{0.5bx}{A_s}$$

From (2)

$$f_s/f_{cb} = \frac{E_s}{E_c} \cdot \frac{d-x}{x} = \frac{\alpha_e(d-x)}{x}$$

$$\therefore \quad \frac{0.5bx}{A_s} = \frac{\alpha_e(d-x)}{x}$$

or

$$0.5bx^2 = \alpha_e A_s(d-x) \tag{3}$$

Writing $\rho = \dfrac{A_s}{bd}$ and $x = x_1 d$

$$0.5x_1^2 bd^2 = \alpha_e \rho bd^2(1-x_1)$$

or

$$x_1^2 = 2\alpha_e \rho(1-x_1)$$

and
$$x_1 = \alpha_e \rho \left(\sqrt{1 + \frac{2}{\alpha_e \rho}} - 1 \right)$$

The value of the modular ratio $\alpha_e = E_s/E_c$ may be taken as 15.

$$\therefore \quad x_1 = 15\rho \left(\sqrt{1 + \frac{2}{15\rho}} - 1 \right) \tag{4}$$

Solving equation (4) gives x_1.

Lever arm factor $z_1 = 1 - x_1/2$.

The moment of resistance of the section based on the steel stress is given by

$$M_r = A_s f_s z$$

or
$$\frac{M_r}{bd^2 f_s} = \rho z_1 = \rho(1 - x_1/3) \tag{5}$$

This ratio is plotted in figure 3.11.

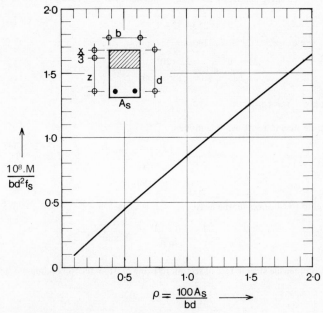

Figure 3.11 Elastic design chart—cracked section (b and d in mm)
f_{st} = stress in tension reinforcement (N/mm^2)
M = applied moment (kN m)
α_e = 15 = modular ratio
A_s = area of tension steel (mm^2)

3.6.2   Tension reinforcement

Assuming equal steel in each face of the section, the area of steel required to resist a structural tensile service load of T at a service stress of f_s is given by

$$A_s = \frac{T}{2f_s} \tag{6}$$

The value of f_s used in equation (6) must be identical to the value used in equation (5).

Example 3.4

Limiting stress design
Calculate the necessary reinforcement in a wall panel subject to a bending moment of 35 kN m/m together with a direct tensile force of 50 kN/m.

Wall thickness $h = 250\,$mm   Cover $c = 52\,$mm
Concrete grade 25   Steel $f_y = 425\,$N/mm^2
Exposure class A

Allowable tensile stress in steel $= 100\,$N/mm^2

$$\text{Effective depth } d = 250 - 52 - \frac{\phi}{2} \text{ (assume bar size 16 mm)}$$

$$= 250 - 52 - 8$$

$$= 190\,\text{mm}$$

$$\frac{M \times 10^8}{f_{st}bd^2} = \frac{35 \times 10^9}{100 \times 10^3 \times 190^2} = 0.97$$

From graph $\rho = 1.14$

Area of tensile steel to resist bending monent

$$A_s = \frac{1.14 \times 10^3 \times 190}{100}$$

$$= 2160\,\text{mm}^2$$

Area of steel to resist direct tensile force

$$= \frac{T}{f_{st}} = \frac{50 \times 10^3}{100} = 500\,\text{mm}^2$$

∴ Total area of tensile steel in face of section resisting tension due to bending moment

$$A_1 = 2160 + \tfrac{1}{2} \times 500$$

$$= 2410\,\text{mm}^2$$

On opposite face

$$A_2 = \tfrac{1}{2} \times 500 = 250\,\text{mm}^2$$

Minimum steel required to resist early thermal movement is greater than this value—say 0.15% each face.

$$\therefore A_2 = 0.15\% \times 1000 \times 250$$

$$= 375\,\text{mm}^2$$

Face 1 use Y20 at 125 (2510)
Face 2 use Y12 at 250 (452)

3.7   Design for no cracking

The calculation of crack widths under flexural loading is a relatively recent innovation, and it was previously usual to design on the theoretical basis of 'no cracking' under service loads, assuming an uncracked concrete section and limiting the tensile stress in the concrete. The method was in use for many years, but is not as economical as the limit state method previously described. Concrete sections designed by this method tend to be thick and have relatively large amounts of reinforcement. In BS 5337[10] this design method is referred to as the Alternative Method of Design. The fundamental assumptions are:

1. In calculations relating to resistance to cracking, the concrete is assumed to be capable of resisting a limited tensile stress and the whole section, including cover to the reinforcement, is taken into account.
2. In strength calculations, it is assumed that the concrete has no tensile strength. The necessary formulae may be developed from elastic design theory taking into account the strength of the concrete in tension.

3.7.1   *Resistance to cracking*
The section to be analysed is shown in figure 3.12. The depth of the neutral axis x is determined by equating the forces of tension in the concrete and steel with the force of compression in the concrete.

Tensile force in the steel $= A_s f_s$
Tensile force in the concrete $= \frac{1}{2} f_{ct} b (h - x)$
Compressive force in the concrete $= \frac{1}{2} f_c b x$

$$\therefore A_s f_s + \tfrac{1}{2} f_{ct} b (h - x) = \tfrac{1}{2} f_c b x \qquad (1)$$

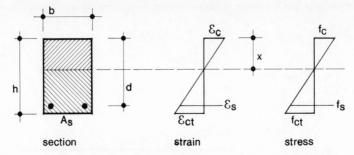

Figure 3.12   Assumed stress and strain diagrams—uncracked section—elastic design

Also from the geometry of the strain diagram and puttting the modular ratio $\alpha_e = E_s/E_c$

$$\frac{\varepsilon_c}{x} = \frac{\varepsilon_s}{d-x} = \frac{\varepsilon_{ct}}{h-x} \qquad (2)$$

and

$$\frac{f_c}{x} = \frac{f_s}{\alpha_e(d-x)} = \frac{f_{ct}}{h-x}. \qquad (3)$$

Substituting in (1)

$$\frac{A_s f_{ct} \alpha_e (d-x)}{(h-x)} + \tfrac{1}{2} f_{ct} b (h-x) = \frac{\tfrac{1}{2} f_{ct} b x^2}{(h-x)}$$

or

$$A_s \alpha_e (d-x) + \tfrac{1}{2} b (h-x)^2 = \tfrac{1}{2} b x^2$$

If

$$\rho_c = \frac{A_s}{bh} \qquad (4)$$

$x = x_1 h$ and $\alpha = d_1 h$

$$x_1 = \frac{0.5 + \rho_c \alpha_e d_1}{1 + \rho_c \alpha_e} \qquad (5)$$

The value of the modular ratio α_e may either be taken to allow for the concrete which is displaced by the reinforcement or this small difference may be ignored. A value of $\alpha_e = 15$ is sufficiently accurate for most conditions.

The applied moment may be equated to the moment of the tensile forces in the steel and concrete about the centre of gravity of the compressive force in the concrete. This is at a distance $x_1 d/3$ from the compression face of the section. Hence

$$M = A_s f_s \left(d - \frac{x}{3} \right) + \tfrac{1}{2} f_{ct} b (h - x) \tfrac{2}{3} \tag{6}$$

From (3), (4) and (6),

$$M = \rho_c bh f_{ct} \frac{\alpha_e (d-x)}{h-x} \left(d - \frac{x}{3} \right) + \tfrac{1}{3} f_{ct} b (h-x) h$$

or

$$\frac{M}{bh^2 f_{ct}} = \frac{(1-x_1)^2 + \rho_c \alpha_e (d_1 - x_1)(3d_1 - x_1)}{3(1-x_1)} \tag{7}$$

Values of the moment factor from equation (7) are plotted in figure 3.13.

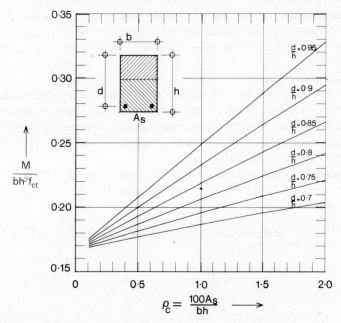

Figure 3.13 Elastic design chart for uncracked section (b, d, and h in mm)
f_{ct} = tensile stress in concrete (N/mm²)
M = applied moment (kN m)
$\alpha_e = 15$ = modular ratio
A_s = area of tension steel (mm²)

Table 3.6 Permissible concrete stresses in calculations relating to the resistance to cracking for reinforced concrete

	Permissible concrete stresses in N/mm^2	
		Tension
Concrete grade	Direct	Due to bending
Grade 30	1.44	2.02
Grade 25	1.31	1.84

The permissible values of tensile stress in the concrete are given in table 3.6 (from BS 5337).

3.7.2 Strength

The calculation to determine the strength of the section is made by elastic theory at service loads, assuming that the concrete has cracked in tension. This assumption is at variance with the assumption made in section 3.7.1 for controlling cracking and can only be substantiated on the basis that *if* the section cracks, the reinforcement will still be adequate to prevent failure.

The equations used are identical with those used in section 3.6. The allowable stresses in the reinforcement depend on the Code of Practice being followed, and the values taken from BS 5337[10] are given in table 3.7.

3.7.3 Section thickness

The section thickness may be calculated by considering the moment of resistance of an *uncracked* section assisted by an assumed quantity of reinforcement. The moment of resistance of such a section depends on the overall thickness, the ratio of effective depth to overall depth, and the permissible tensile stress in the concrete.

The quantity of reinforcement may be assumed to be 0.5% in order to arrive at a section thickness, and checked after the steel required has been more accurately calculated.

Table 3.7 Permissible steel stresses in strength calculations

Condition	Class of exposure	Permissible stress in N/mm^2 (Deformed bars)
Direct tension	A	100
Flexural tension		
Shear	B	130
Compression	A and B	140

Table 3.8 Values of factor K (N/mm^2) in formula
$M_r = Kbh^2$. Uncracked section
$f_{ct} = 1.84\,N/mm^2$. $\alpha_e = 15$

$\dfrac{100A_s}{bh}$	0	0.5	0.75	1.00	1.50
$\dfrac{d}{h}$					
0.75	0.30	0.33	0.34	0.36	0.38
0.80	0.30	0.34	0.36	0.38	0.41
0.85	0.30	0.35	0.38	0.40	0.44
0.90	0.30	0.37	0.40	0.43	0.49

The moment of resistance is given by

$$M_r = Kbh^2$$

where K has a value taken from table 3.8, and the applied moments are calculated using the service loads. The section thickness is determined to ensure that the moment of resistance of the section is greater than the applied moment.

3.7.4 Calculation of reinforcement
The choice of the section thickness should ensure that there are no flexural cracks in the section under service loads, but the quantity of reinforcement is decided by assuming a cracked section as shown in figure 3.10 and providing steel at a low stress to resist the tension due to flexural action. The permissible stresses are given in table 3.7.

3.7.5 Combined flexure and tension
When a section is subjected to both direct tension and flexural action, some adjustment has to be made as the permissible stresses for each type of action are different. The calculations are conveniently carried out using an equivalent transformed section, i.e. considering the whole section in terms of concrete with the steel contributing at a ratio of α_e times its own area.

The maximum tensile stress on the whole transformed section due to the applied moment is

$$f_m = \frac{M(d-hx)}{I_e}$$

and the stress due to the direct force is

$$f_d = \frac{F}{bh}$$

The section should be proportioned such that

$$\frac{f_m}{f_{mp}} + \frac{f_d}{f_{dp}} \leqslant 1.0$$

where h_x is the depth to the neutral axis
          I_e is the second moment of area of the transformed section
          f_{mp} is the permissible stress due to bending
          f_{dp} is the permissible stress due to direct tension
A typical calculation is given in Example 3.5.

Example 3.5

Alternative method of design
Design a section to resist an applied bending moment of $M = 90\,\text{kN m/m}$ in addition to a direct force $F = 75\,\text{kN/m}$. Class B exposure. Grade 25 concrete. High-yield steel.

(1) *Section thickness*
Assume 0.5% reinforcement and $\dfrac{d}{h} = 0.9$.

If only the moment applied, from table 3.8, $K = 0.37$

$$\therefore \ M = Kbh^2$$

or $90 \times 10^6 = 0.37 \times 10^3 \times h^2$

$$\therefore \ h = 487\,\text{mm}$$

With a tensile applied force in addition, some extra thickness will be required.
  Try

$$h = 550\,\text{mm}$$

$$M_r = 0.37 \times 10^3 \times 550^2 \times 10^{-6} = 112\,\text{kN m/m}$$

$$\frac{f_m}{p_m} = \frac{1.84 \times \frac{90}{112}}{1.84} = 0.80$$

For tensile force

$$f_t = \frac{75 \times 10^3}{550 \times 10^3} = 0.14\,\text{N/mm}^2$$

$$p_t = 1.31$$

$$\therefore \ \frac{f_t}{p_t} = \frac{0.14}{1.31} = 0.11$$

$$\therefore \ \frac{f_m}{f_{mp}} + \frac{f_t}{f_{tp}} = 0.80 + 0.11 = 0.91$$

which is < 1.0 and $\therefore$ satisfactory

(2) *Strength*
  Concrete to grade 25. $p_{cb} = 9.15\,\text{N/mm}^2$
  High yield deformed bars $p_{st} = 130$ (class B exposure)
  Modular ratio $\alpha_e = 15$.   Cover $= 40\,\text{mm}$
  Axial distance $= 40 + 10 = 50\,\text{mm}$

From previous calculation $h = 550\,\text{mm}$, $\therefore d = 550 - 50 = 500\,\text{mm}$. The depth of the neutral axis under flexure only is given by

$$x = 0.26d = 0.26 \times 500 = 130\,\text{mm}$$

$$\text{Lever arm} = \left(1 - \frac{0.26}{3}\right) \times 500 = 0.91 \times 500 = 455\,\text{mm}$$

$$\therefore A_{st} = \frac{M}{f_{st}z} = \frac{90 \times 10^6}{130 \times 455} = 1522\,\text{mm}^2$$

check

$$\rho_c = \frac{1522}{10^3 \times 550} = 0.28\%$$

$$\frac{d}{h} = \frac{500}{550} = 0.91$$

$\therefore$ assumptions in (1) above are satisfactory as actual ρ_c is $<$ assumed value and actual $\dfrac{h}{h}$ is $>$ assumed value.

Reinforcement required for resistance to tensile force

$$= \frac{F}{p_{st}} = \frac{75 \times 10^3}{130} = 577\,\text{mm}^2 \text{ provided in each face equally.}$$

$$\therefore \text{ Total } A_s = 1522 + \frac{577}{2} = 1811\,\text{mm}^2$$

$$A_s' = \frac{577}{2} = 290\,\text{mm}^2$$

Minimum reinforcement in each face may be calculated or nominal (BS 5337, 4.11). Assume that 0.3% total is required. i.e. 0.15% in each face

$$0.15\% \times bh = 0.15\% \times 10^3 \times 550$$

$$= 825\,\text{mm}^2$$

$\therefore$ Provide:

'Tension' face 1811 mm^2: $\underline{\text{Y20 at 150 (2090)}}$

'Compression' face 825 mm^2: $\underline{\text{Y16 at 200 (1010)}}$

with $h = 550$.

3.8 Bond and anchorage

At the overlap of bars transmitting tension, it is preferable to be generous with the length of overlap to avoid cracking at the ends of the lapped bars.

Distribution reinforcement or any reinforcement acting to resist early thermal stresses should be designed to have lap lengths sufficient to resist the yield or proof strength of the bar (table 3.9). If a greater steel ratio than required is actually provided, the lap lengths can be reduced in proportion. For limit state design, the required lap lengths are given in table 3.10. Again, it is possible to reduce the laps by the proportion of steel area provided/steel area required. Similar values for the Alternative Method of Design are given in table 3.11, and can also be modified as above[10,6]. The lap lengths in tables 3.9 to 3.11 are based on calculations using the permissible bond stresses and other values in BS 5337 and CP 110.

It is preferable to provide lap lengths at least equal to about 35ϕ, as small lap lengths increase the possibility of cracking at the free ends of the bars, and are proportionately affected more seriously by tolerances and incorrect detailing and fixing.

Table 3.9 Tension lap lengths for the critical reinforcement ratio for shrinkage and thermal movement design

	Plain round mild steel	Deformed high-yield steel (ribbed)
f_y (N/mm²)	250	425
concrete grade		
25	55ϕ	62ϕ
30	48ϕ	55ϕ

Table 3.10 Tension lap lengths for ultimate limit state design

Concrete grade	Plain round mild steel	Deformed high-yield steel (ribbed)
	$f_y = 250$	$f_y = 425$
25	39ϕ	47ϕ
30	37ϕ	41ϕ

Table 3.11 Tension lap lengths for alternative design

	Plain round mild steel		Deformed high-yield steel (ribbed)	
f_{dst} (N/mm²)	85	115	100	130
concrete grade				
25	24ϕ	32ϕ	20ϕ	26ϕ
30	22ϕ	29ϕ	18ϕ	24ϕ

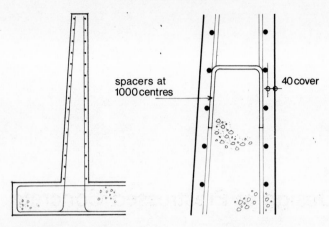

spacers at
1000 centres

40 cover

Figure 3.14   Detailing of spacer reinforcement

3.9   Detailing[27,28,29]

The detailing requirements for water-retaining structures follow the usual
rules for normal structures. Bars should be detailed for continuity on the
liquid faces and sudden changes of reinforcement ratio should be avoided.
The distribution reinforcement in walls should be placed in the outer layers
where it has maximum effect. Spacers should be detailed to ensure that the
correct cover is maintained (figure 3.14).

Welded fabric reinforcement is normally used to reinforce floor slab
panels and may also be used in some walls where the required
reinforcement area is not too large.

4   Design of Prestressed Concrete

4.1   The use of prestressed concrete

Prestressed concrete is a structural material in which compressive stresses are induced in the concrete before imposed loading is applied. The magnitude of the induced stresses is arranged so that, after the application of imposed loads, the stresses in the concrete are still largely compressive. Prestressing can be applied in a slab in one direction or in two orthogonal directions in the plane of the slab. Prestressed concrete is divided into two types.

1. Pre-tensioned: in which long wires are stretched on a tensioning bed in the factory. Concrete is then placed in moulds around the wires which are released when the concrete is hardened. The wires are then cut at intervals along the length to create separate elements.
2. Post-tensioned: in which concrete elements are cast in place and subsequently stressed with external or internal wires.

Prestressed concrete would appear to have a considerable advantage for use in liquid-retaining structures in that the concrete is in compression, there are no cracks, and leakage is not possible. In practice, it is difficult to make use of this advantage, but various applications are discussed in the following sections.

It is not possible in a book of this length to deal fully with the theory and practice of prestressed concrete, and for further information the reader is referred to *Design of Prestressed Concrete* by Bate and Bennett[30].

4.2   Materials

4.2.1   *Concrete*
The maximum stresses on the concrete are not usually very high in relation to the strength of the concrete. A strength of $40 \, \text{N/mm}^2$ will generally be

satisfactory and provide sufficient durability. It is important to ensure adequate workability in order to achieve full compaction.

4.2.2 Prestressing tendons[18,31]

Prestressing wires or strands of the normal commercial grades may be used. The jacking force is limited to 70% of the characteristic ultimate strength of the tendon, and losses due to friction, slip of the grips, relaxation of the steel, and elastic contraction, shrinkage and creep of the concrete must be considered when arriving at the final active prestressing forces. Tendons may be drawn through ducts cast into the concrete section or placed on the outside of a wall (and subsequently covered with pneumatically placed concrete to give cover to the steel to ensure durability).

A proprietary system of winding wire under tension around circular tanks is widely used.

Frictional losses in post-tensioned prestressed concrete can occur between the cables and the sides of the ducts, caused by the unintentional curvature of the duct, and the frictional forces will be increased by the duct not being exactly in line, and having irregularities in the profile. To allow for the curvature, the loss may be calculated from

$$P_0(1 - e^{-\mu x/R})$$

To allow for irregularities, the loss may be taken as

$$P_0(1 - e^{-kx})$$

where
P_0 = prestressing force at the jack
e = 2.718
R = radius of curvature of the duct
μ = coefficient of friction
$\quad \mu = 0.55$ for steel on concrete
$\quad \mu = 0.30$ for steel on steel
k = a constant depending on the type of duct which may vary between 33×10^{-4} and 17×10^{-4}.

4.3 Precast prestressed elements

It is not generally possible to make use of precast prestressed wall elements because of the difficulty in making an economic connection between the units longitudinally and also between the wall units and the floor. However, the use of a precast prestressed roof to a reservoir or tank may be economical. Reservoir roofs are generally supported by reinforced concrete columns at centres of about 5 to 7 metres. Where climatic conditions require speedy construction due to approaching bad weather, a precast design enables the construction period to be shortened.

The design of roof elements follows the normal principles of prestressed concrete design. Because of condensation within a confined space, the underside of the roof will normally be rated as exposure type A (see chapter 2) requiring the units to be designed to have no tensile stresses on the underside with full imposed loading. The imposed loads may include soil cover and vehicles during construction.

4.4 Cylindrical prestressed concrete tanks[32,33]

4.4.1 *Loads and horizontal forces*

A cylindrical tank (with a vertical axis) is a convenient structure to contain liquid. The radial pressure due to the liquid is uniform at all points on the circumference at a given depth. Each circular slice of the tank wall at a given level is therefore in equilibrium under the applied liquid pressure, and horizontal ring tensile forces are developed in the wall. The pressure and forces vary linearly with depth from zero at the liquid surface to a maximum at the tank floor, and hence the induced horizontal ring tension also varies. In reinforced concrete construction, the wall deflects outwards by an amount which varies with depth (ignoring any effect of the floor) (figure 4.1).

The use of prestressed concrete enables compressive ring forces to be induced in the wall which counteract the tensile forces due to the liquid. Stressing tendons may be placed more closely in the lower part of the wall and more widely spaced in the upper sections. Thus, in theory, it is possible to arrange for zero stresses in the concrete with a tank full of liquid. In practice, the variation in spacing of the wires will not be continuously

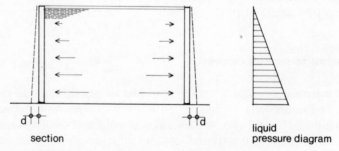

section   liquid pressure diagram

Figure 4.1 Outward deflection of cylindrical tank (free at base)

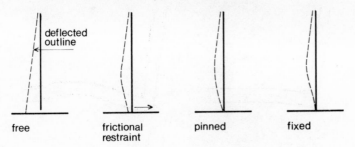

Figure 4.2   Effect of base restraint on a loaded prestressed concrete tank

variable and, in order to avoid leakage, it is desirable to have a larger prestressing force throughout the wall height than is necessary to counteract exactly the applied loading, so that there is always a residual compressive stress in the concrete. A value of $1.0 \, \text{N/mm}^2$ is reasonable after allowing for all losses of prestress.

4.4.2   Base restraint

The discussion in section 4.4.1 has ignored the effect of the connection between the wall and the floor. There are three possible types of connection (figure 4.2):

(a) fixed
(b) pinned
(c) free to slide

It is clear that if a fixed joint is used, it is not possible to prestress the concrete wall effectively near to the base, as no inward radial movement is possible.

A nominally free joint provides the least resistance to circumferential prestress, but in practice it is not possible to avoid frictional forces between the wall and base due to the deadweight of the wall. In view of the doubt about the extent of any restraint, the circumferential prestressing should be designed on the basis of no restraint. This is a safe procedure. The effect of restraint is to cause a varying vertical bending moment in the wall section.

The radial deflection of the wall under load is shown in figure 4.2 for the various forms of restraint.

Fixed restraint has been shown to be disadvantageous when considering the circumferential prestress, and is also difficult to achieve. The tank base is founded on soil which will deflect under the weight of the tank and the moment due to fixity, and the base itself is to some degree flexible (figure 4.3).

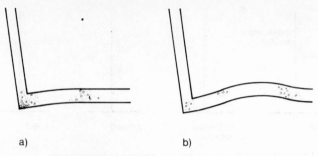

a)   b)

Figure 4.3   Rotation of a "fixed" wall footing
  (*a*) Due to settlement
  (*b*) Due to flexibility of the footing

A compromise is possible between the three types of fixity which is convenient in practice. The joint is made nominally free to slide during the prestressing operation, and then pinned in position so that under full-load conditions the joint acts as a pin.

4.4.3   *Vertical design*
Reinforcement must be provided in the vertical direction to resist the following forces:

  (*a*) Bending moments induced by the variation of prestress with depth when the tank is empty.
  (*b*) Bending moments induced due to base fixity.
  (*c*) Load out-of-balance moments created during the prestressing operation.
  (*d*) Bending moments due to variation in temperature between the outside and inside of the tank.

(*a*) and (*b*) have been discussed already. (*d*) is of importance only in countries where the sun can cause a high surface temperature on the concrete. In temperate climates it is not usual to take account of this effect.

Vertical bending moments may be resisted by normal reinforcement or by further vertical prestressing or a combination of both. For tanks up to about 7 m deep, it will be more economical to use normal reinforcement[32].

5   Distribution Reinforcements and Joints: Design Against Shrinkage and Thermal Stresses

Cracks are induced in reinforced concrete members by the action of applied loads and by environmental conditions. This chapter considers the effect of environmental conditions on concrete slabs which are assumed to be otherwise unloaded[9,34]. A typical practical example is in the longitudinal direction of a cantilever reservoir wall.

5.1   Cracking in reinforced concrete

If a small cube of concrete is cast (such as a test cube), it will not exhibit cracking apparent to the naked eye. However, if a long specimen with relatively small cross-sectional dimensions is made containing reinforcement, and the ends of the reinforcement are held to prevent any movement, after a few days it will be found that fine lateral cracks are present (figure 5.1). Depending on the relation between the quantity of reinforcement, the bar size, and the cross-sectional area of concrete, either a few wide cracks, or a larger number of fine cracks will form. The cracks are induced by the resistance of the reinforcement to the strains in the concrete caused by chemical hydration of the cement in the concrete mix[35].

5.2   Causes of cracking

5.2.1   *Heat of hydration*
When concrete materials are mixed together, a chemical reaction takes place between the cement and water, during which heat is evolved. This

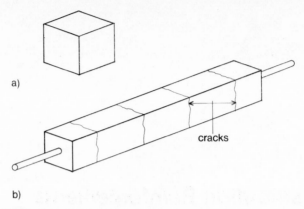

Figure 5.1 Cracking in concrete elements
(a) Small cube free to move
(b) Long reinforced element restrained by reinforcement

heat of hydration causes the temperature of the concrete to rise until the reaction is complete, and the heat is then dissipated to the atmosphere. A typical curve illustrating the temperature rise in concrete during the first few days after mixing is shown in figure 5.2a. By the sixth day, the temperature is usually back to normal. The value of the maximum temperature is dependent on the quantity of cement in the mix, the thickness of the concrete section, and any insulation that is provided, deliberately, or by formwork. Concretes which are rich in cement will emit larger quantities of heat than concretes with low cement content[36]. A thick concrete section (over about 800 mm) will not cool very quickly, because the ratio of surface area to total heat emitted is lower. Recent work on the avoidance of cracking has shown that it may be advantageous to allow thick sections to cool slowly by preventing rapid loss of heat. This is achieved by covering the exposed concrete with an insulating blanket. For normal structural work, the formwork should not be removed for three or four days, otherwise cold winds may cause surface cracking of the warm concrete[37,38].

During the period when the concrete temperature is increasing, expansion will take place. If the expansion is restrained by adjoining sections of hardened concrete, some creep will occur in the relatively weak concrete, relieving the compressive stresses induced by the attempted expansion. As the concrete subsequently cools, it tries to shorten but, if there are restraints present, tensile strains will develop leading to cracking (figure 5.2b).

5.2.2 Drying shrinkage
As concrete hardens and dries out, it shrinks. This is an irreversible process.

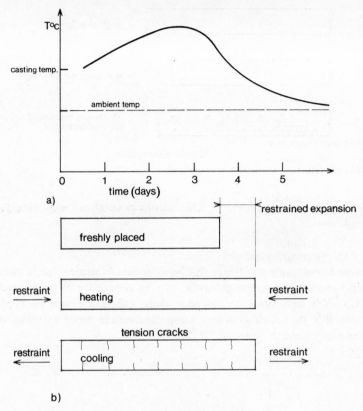

Figure 5.2   (*a*) Rise in temperature of freshly placed concrete
(*b*) Thermal strains

If a reinforced concrete member is considered (figure 5.3) under no external stress, it will be apparent that free shrinkage of the concrete is prevented by the steel reinforcement. The steel is therefore in compression and the concrete in tension, with longitudinal bond forces present on the surface of the reinforcement. The magnitude of these forces is dependent on the concrete properties, and the ratio of the area of steel to the area of concrete. If a high ratio of steel is present, and there is no external restraint applied to the element, no cracks will form but, if the steel ratio is relatively small or external restraints are present, cracking is certain. The cracks may form at close centres and be fine in width, or may be further apart and be correspondingly wider in order to accommodate the total strain. It is important that there is sufficient reinforcement to control the cracking. If this is not the case, a few very wide cracks will form, and the reinforcement

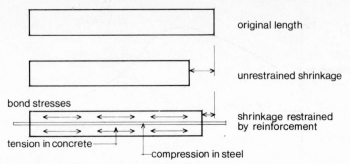

Figure 5.3 Drying shrinkage in reinforced concrete

will yield at the crack positions[9]. The various possibilities are illustrated in figure 5.4.

5.2.3 Environmental conditions

An elevated concrete water tower will be subjected to strains due to changes between summer and winter temperatures. In temperate climates (such as the UK), it is not usual to consider these effects for normal types of structures, but in countries where temperatures are more extreme, some allowance may need to be made[39].

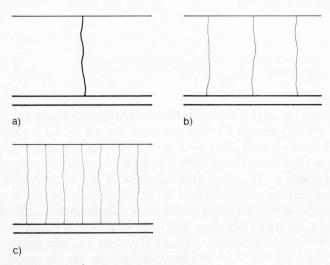

Figure 5.4 Development of cracks related to reinforcement
  (a) Insufficient steel—wide cracks, steel yields
  (b) Controlled cracks—average spacing and width
  (c) High proportion of steel—well-controlled cracks of narrow width and close spacing

The effect of the sun in heating part of the surface of a structure may produce differential strains between one side and another. Again, in temperate climates, these are usually ignored in calculation, but in hot countries they will have to be considered[39].

5.3 Crack distribution[34]

Cracking due to early thermal movement may be controlled by reinforcement (figure 5.5). The objective is to distribute the overall strain in the wall by reinforcement and by movement joints, so that the crack widths are acceptable or, if considered desirable, that the concrete remains uncracked. There is no single solution to the design problem of controlling early thermal cracking. The designer may choose to have closely spaced movement joints with a low ratio of reinforcement, or widely spaced joints with a high ratio of reinforcement. The decision is dependent on the size of the structure, method of construction to be adopted, and economics.

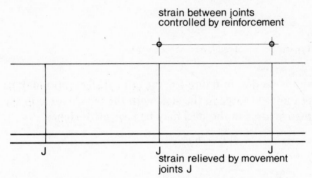

Figure 5.5 Relation of movement joints and reinforcement in controlling strain

5.3.1 *Critical steel ratio*
Figure 5.6 shows part of a reinforced concrete slab. The concrete section is taken as an area on each side of the bar, corresponding to half the distance to the next reinforcing bar. Alternatively, a unit width of one metre may be defined, and the total area of steel within this width may be read from bar spacing tables.

The steel ratio in a section is defined as the ratio

$$\rho_c = \frac{A_s}{bh}$$

and is often expressed as a percentage, where

A_s is the total area of reinforcement (in both faces)
b is the width of the section
h is the overall thickness of the section.

If a section contains a low steel ratio, the strength of the steel at yield will be less than the ultimate concrete strength in tension. Cracks will be unrestrained as the steel yields.

At a certain critical steel ratio, the reinforcement yields and the concrete reaches its ultimate tensile stress at the same time. If the section is reinforced, so that the actual steel ratio is not less than the critical ratio, then any cracks which form will be restrained and of moderate width.

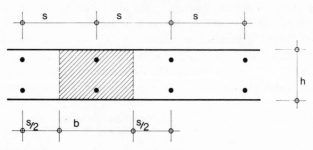

Figure 5.6   Typical section of reinforced concrete slab

Considering the section in figure 5.6, the critical steel ratio may be obtained by equating the yield force in the steel with the tensile force in the concrete. The compressive force in the steel may be neglected. Hence

$$A_s f_y = bh f_{ct}$$

and

$$\left(\frac{A_s}{bh}\right)_{crit} = \rho_{crit} = \frac{f_{ct}}{f_y}$$

The control of cracking is critical during the early life of the concrete, and therefore a value of concrete tensile strength at 3 days may be used. Typical values of ρ_{crit} and f_{ct} are given in table 5.1.

5.3.2   Crack spacing

If sufficient reinforcement is provided to control the cracking ($\rho_c > \rho_{crit}$), then the probable spacing of the cracks may be estimated. Cracks form in sequence, as the shrinkage strain increases when the bond force between the reinforcement and the concrete becomes greater than the tensile strength of the concrete. The bond stress between concrete and steel that accompanies

Table 5.1 Values of critical steel ratios

Concrete grade (N/mm^2)	f_{ct} (N/mm^2)	f_y (N/mm^2)	ρ_{crit} %
30	1.3	425	0.31
		250	0.52
25	1.15	425	0.27
		250	0.46

f_{ct} = direct tensile strength of the concrete at 3 days.
f_y = characteristic yield strength of reinforcement.
ρ_{crit} = critical steel ratio (the minimum required to control cracking in a restrained section)

the formation of a crack extends for a length equal to half the crack spacing (figure 5.7). Equating the two forces gives

$$f_b s \Sigma u_s = f_{ct} bh \tag{1}$$

where

Σu_s = total perimeter of bars
f_b = average bond stress adjacent to a crack
s = bond length necessary to develop cracking force
f_{ct} = tensile stress in concrete
bh = area of concrete

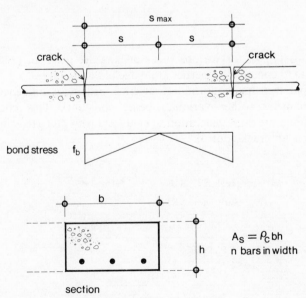

$$A_s = \rho_c bh$$
n bars in width

Figure 5.7 Bond stress and crack formation

Writing steel ratio $\rho_c = \dfrac{A_s}{bh}$ (neglecting concrete area taken up by steel) and the ratio

$$\frac{\text{total perimeters}}{\text{total steel area}} = \frac{\Sigma u_s}{A_s}$$

$$= \frac{\pi\phi \times (\text{number of bars})}{\dfrac{\pi}{4}\phi^2 \times (\text{number of bars})} = \frac{4}{\phi}$$

where ϕ = bar size (or equivalent size for square or ribbed bars), substitution in (1) gives

$$f_b s\left(\frac{4}{\phi} A_s\right) = f_{ct} bh$$

$$\therefore \; s = \left(\frac{f_{ct}}{f_b}\right)\left(\frac{1}{\rho_c}\right)\left(\frac{\phi}{4}\right)$$

$$= \left(\frac{f_{ct}}{f_b}\right)\frac{\phi}{4\rho_c}$$

and the maximum crack spacing

$$s_{max} = 2s = \left(\frac{f_{ct}}{f_b}\right)\frac{\phi}{2\rho_c}$$

The ratios (f_{ct}/f_b) for various types of bars are given in table 5.2. The ratios relate to concrete properties at an age of about 3 days.

It is apparent from the formula that crack spacing is influenced directly by bar diameter (other variables being constant). This confirms the judgement of a previous generation of engineers who preferred small bars at close centres for crack control.

Table 5.2   Ratios of $\left(\dfrac{f_{ct}}{f_b}\right)$ at early ages

Bar type	Ratio
Plain bars	1
Deformed bars type 1	4/5
Deformed bars type 2	2/3

For a given steel ratio, it is possible to choose the bar size (within limits) so that the crack spacing is small, and the total contraction strain is accommodated by the formation of many fine cracks, or by choosing a larger bar size, to cause cracks to form at wider centres. If joints are placed at the assumed centres of crack formation, the concrete will effectively be uncracked.

5.3.3 Crack widths

The number and width of cracks which form will depend on the total contraction strain which is unrelieved by joints in the length of the section. The contraction strain is the sum of the shrinkage strain and the thermal strain (due to changes in the ambient temperatures after the structure is complete). Assuming that the steel ratio is greater than ρ_{crit} and with full restraint (i.e. no joints), the tensile strain in the concrete may be assumed to vary from zero adjacent to a crack to a value of ε_{ult} (ultimate concrete strain) midway between cracks at a distance s_{max} apart. The average tensile strain in the uncracked concrete is therefore $\frac{1}{2}\varepsilon_{ult}$. The strain due to the maximum crack width, which is the difference between the total contraction strain and the strain remaining in the concrete, is therefore given by

$$\left(\frac{w}{s_{max}}\right) = \varepsilon_{te} + \varepsilon_{cs} - \tfrac{1}{2}\varepsilon_{ult}$$

where w = maximum crack width
     s_{max} = maximum spacing of cracks
     ε_{cs} = total shrinkage strain
     ε_{ult} = ultimate concrete tensile strain
     ε_{te} = thermal contraction from peak temperature.

There is no sufficient information available to enable precise values for the various coefficients to be given, but ε_{ult} may be assumed to be 200 microstrains. The shrinkage strain in the concrete, less creep strain, is about 100 microstrains and therefore in the formula above equates with the value of $\frac{1}{2}\varepsilon_{ult}$. The remaining strain to be considered, ε_{te}, is therefore due to cooling from the peak of hydration temperature T_1 to ambient temperature. There is also a further variation in temperature T_2 due to seasonal changes after the concrete in the structure has hardened.

When considering the strain due to temperature T_1, an effective coefficient of expansion of one half of the value for mature concrete should be used due to the high creep strain in immature concrete. For mature concrete and seasonal variations due to temperature T_2, the tensile strength of the concrete is lower compared with the bond strength, hence s is much less for mature concrete when T_2 is appropriate; hence the actual contraction can be effectively halved. The strain equation now becomes

$$\left(\frac{w}{s_{max}}\right) = \tfrac{1}{2}\alpha(T_1 + T_2) + 100 - (\tfrac{1}{2} \times 200)$$

$$= \tfrac{1}{2}\alpha(T_1 + T_2)$$

$\therefore$ Crack width $= w = s_{max}\tfrac{1}{2}\alpha(T_1 + T_2)$

where $\alpha =$ coefficient of linear expansion of concrete.

Typical values for T_1 and T_2 applicable to conditions in the UK are given in table 5.3. The values for T_1 may need to be increased if the cement content is over $340\,kg/m^3$ or if the wall thickness exceeds 400 mm.

Table 5.3 Values of T_1 and T_2

	Temperature	Strain ($\tfrac{1}{2}\alpha T$)
T_1		
concreting in summer	30°C	180
concreting in winter	20°C	120
T_2		
summer construction	20°C	120
winter construction	10°C	60

$\alpha = 12$ microstrains/°C
Note: See section 5.4 for reduction in T_2.

The theory given in this section has been developed by Professor B. P. Hughes, University of Birmingham[9].

5.4 Joints[40,41]

5.4.1 *Construction joints*

It is rarely possible to build a reinforced concrete structure in one piece. It is therefore necessary to design and locate joints which allow the contractor to construct the elements of the structure in convenient sections. In normal structures, the position of the construction joints is specified in general terms by the designer, and the contractor decides on the number of joints and their precise location subject to final approval by the designer.

In liquid-retaining structures this approach is not satisfactory. The design of the structure against early thermal movement and shrinkage is closely allied to the frequency and spacing of all types of joints, and it is essential for the designer to specify on the drawings exactly where construction joints will be located. Construction joints are specified where convenient breaks in placing concrete are required. Concrete is placed separately on either side of a construction joint, but the reinforcement is continuous through the joint. At a horizontal construction joint, the free surface of the concrete must be finished to a compacted level surface. At the junction between a base slab and a wall, it is convenient to provide a short 'kicker' which enables the formwork for the walls to be placed accurately and easily. A vertical joint is made with formwork. Details are shown in figure 5.8.

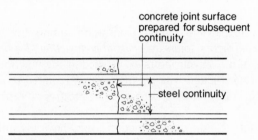

Figure 5.8 Construction joint

Construction joints are not intended to accommodate movement across the joint but, due to the discontinuity of the concrete, some slight shrinkage may occur. This is reduced by proper preparation of the face of the first-placed section of concrete to encourage adhesion between the two concrete faces. Joint preparation consists in removing the surface laitance from the concrete without disturbing the particles of aggregate. It is preferable to carry out this treatment when the concrete is at least five days old, either by sandblasting or by scabbling with a small air tool. The use of retarders painted on the formwork is not recommended, because of the possibility of contamination of the reinforcement passing through the end formwork. The face of a construction joint is flat. Any shear forces can be transmitted across the joint through the reinforcement. If a construction joint has been properly designed, prepared and constructed, it will retain liquid without a waterstop. Extra protection may be provided by sealing the surface as shown in figure 5.9.

5.4.2 Movement joints

Movement joints are designed to provide a break in the continuity of a slab, so that relative movement may occur across the joint in the longitudinal direction. The joints may provide for the two faces to move apart (contraction joints) or, if an initial gap is created, the joint faces are able to

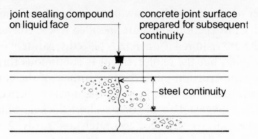

Figure 5.9   Construction joint sealed on the liquid face

move together (expansion joints). Contraction joints are further divided into *complete contraction joints* and *partial contraction joints*.

Other types of movement joints are needed at the junction of a wall and roof slab (figure 5.10), and where a free joint is required to allow sliding to take place at the foot of the wall of a circular prestressed tank (figure 5.11).

Contraction joints. Complete contraction joints have discontinuity of both steel and concrete across the joint, but partial contraction joints have some continuity of reinforcement. Figures 5.12 and 5.13 illustrate the main types. In partial contraction joints, the reinforcement may all continue across the joint, or only 50% of the steel may continue across the joint, the remaining 50% being stopped short of the joint plane. The purpose of the three types of contraction joints is described in section 5.4.1.

Contraction joints may be constructed as such, or may be induced by providing a plane of weakness which causes a crack to form on a preferred line. In this case, the concrete is placed continuously across the joint position, and the action of a device which is inserted across the section, to reduce the depth of concrete locally, causes a crack to form. The formation of the crack releases the stresses in the adjacent concrete, and the joint then

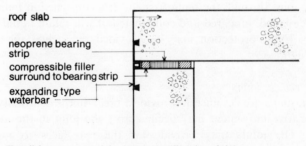

Figure 5.10   Detail for movement joint between wall and roof slab

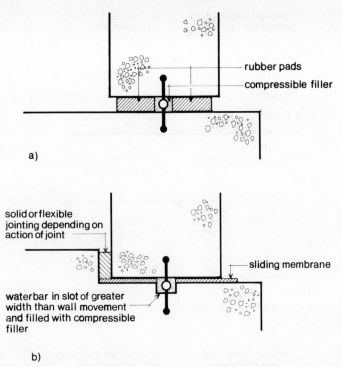

rubber pads

compressible filler

a)

solid or flexible
jointing depending on
action of joint

sliding membrane

waterbar in slot of greater
width than wall movement
and filled with compressible
filler

b)

Figure 5.11 Movement joints between base slab and wall of prestressed concrete tank
(*a*) Rubber pad
(*b*) Sliding membrane

acts as a normal contraction joint. A typical detail is shown in figure 5.14. Great care is necessary to position the crack inducers on the same line, as otherwise the crack may form away from the intended position. Similar details may be used in walls with a circular-section rubber tube placed vertically on the joint-line on the wall centre-line, causing the crack to form.

Expansion joints. Expansion joints are formed with a compressible layer of material between the faces of the joint. The material must be chosen to be durable in wet conditions, non-toxic (for potable water construction), and have the necessary properties to be able to compress by the required amount and to subsequently recover its original thickness. An expansion joint always needs sealing to prevent leakage of liquid. In a wall, a waterbar is necessary containing a bulb near to the centre which will allow movement to take place without tearing (figure 5.15). The joint also requires surface sealing to prevent the ingress of solid particles. By

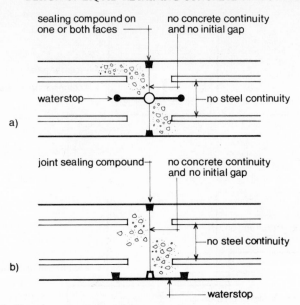

Figure 5.12 Complete contraction joints
  (a) Wall joint
  (b) Floor joint

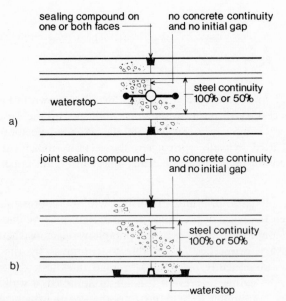

Figure 5.13 Partial contraction joints
  (a) Wall joint
  (b) Floor joint

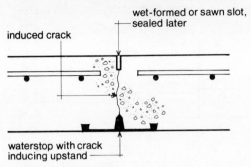

Figure 5.14 Induced contraction joint in floor

definition, it is not possible to transmit longitudinal structural forces across an expansion joint, but the designer may wish to provide for shear forces to be carried across the joint, or to prevent the slabs on each side of the joint moving independently in a lateral direction. If a reservoir wall and footing is founded on ground that is somewhat plastic, the sections of wall on either

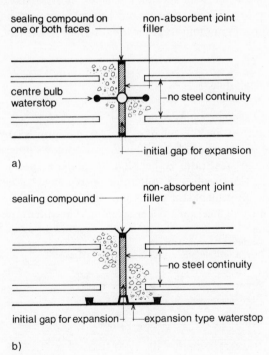

Figure 5.15 Expansion joints
(a) Floor
(b) Wall

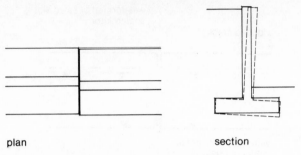

plan   section

Figure 5.16 Lateral movement at unrestrained expansion joint

side of an expansion joint may rotate under load by differing amounts. This action creates an objectionable appearance (figure 5.16). The slabs on either side of an expansion joint may be prevented from relative lateral movement by providing dowel bars with provision for longitudinal movement (a similar arrangement to a road slab). The dowel bars must be located accurately in line (otherwise the joint will not move freely), be provided with an end cap to allow movement, and be coated on one side of the joint with a de-bonding compound to allow longitudinal movement to take place (figure 5.17).

5.5 Typical calculations

5.5.1 *Codes of Practice requirements*
The theory outlined in this chapter has been adopted by the British Standard Code of Practice (BS 5337), which also allows a designer to use a specified minimum steel ratio rather than to prepare specific calculations. The ACI Code[11] specifies a minimum percentage of shrinkage and temperature reinforcement for walls up to 300 mm thick and a fixed minimum quantity of steel for thicker sections. The values are compared in

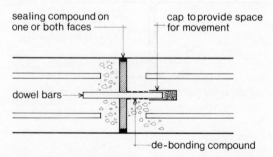

sealing compound on one or both faces   cap to provide space for movement

dowel bars

de-bonding compound

Figure 5.17 Expansion joint including dowel bars to prevent lateral movement

Table 5.4 Specified distribution reinforcement. All tabulated values are for the total minimum amount of reinforcement given as a percentage of the overall cross-sectional area $\quad \rho = \dfrac{A_s}{100bh}$

(a) British Standard 5337:1976	
Plain surface mild steel bars	0.5%
Deformed high yield bars	0.3%
(b) ACI 350R-77	
Walls 300 mm thick:	0.3%
Walls 300 mm thick:	See ACI 318–63[42]

table 5.4 with the critical ratios. For structures of any size, it is recommended that full calculations of the required steel ratio are made.

5.5.2 Calculation of minimum reinforcement

Assume continuous construction, with movement joints at 15 mm centres.

Exposure class B

$$T_1 = 30°C$$

$$T_2 = 0$$

$$\alpha_c = 12$$

Net effective contraction strain $= 0.5 \times 12 \times 30 = 180$ microstrain

Maximum allowable crack width $= 0.2$ mm

$$\therefore \ s_{max} = (0.2/180) \times 10^6 = 1110 \text{ mm}$$

Also
$$s_{max} = \frac{f_{ct}}{f_b} \times \frac{\phi}{2\rho}$$

Assume $\phi = 12$ mm

$$s_{max} = \frac{2}{3} \times \frac{12}{2\rho} = \frac{4}{\rho} \text{ mm}$$

$$\therefore \ \rho = (4/1110 \times 100 = 0.36\%$$

Critical steel ratio $\quad \rho_{crit} = \dfrac{1.3}{425} \times 100 = 0.31\%$

Use $\quad$ <u>Y12 at 150 EF</u> $\qquad$ (1510)

$$\left(\text{Actual ratio} = \frac{1510}{10^3 \times 400} \times 100 = 0.38\% \right)$$

6 Design Calculations

This chapter comprises three design calculations, each forming a complete example and dealing with a particular structure. The handwritten sheets illustrate the graphical method employed in the engineering design process, and each calculation conforms to practical requirements.

6.1 Design of pumphouse

A pumphouse is to be built as part of a sewerage scheme to house three underground electric pumps. The layout is shown in figure 6.1. Design the underground concrete structure. A soil investigation shows dense sand and no ground water.

Design assumptions
The floor and walls must be designed against external soil pressures due to soil and surcharge from vehicles which may park near to the structure. Although no ground water has been found during the site investigation it is quite possible that during the life of the building some ground water may be present on the outside of the walls. Building the structure creates a sump in the original ground which tends to collect water. It is therefore prudent to design the floor and walls to exclude any ground water which may be present. For these reasons, a nominal head of ground water of 1.0 m will be assumed in the structural design. The pump well is designed to hold the effluent as a liquid-retaining structure. Although the normal working level is about mid-height of the walls, it is possible for the effluent to fill the well completely, and for design purposes this condition will be assumed.

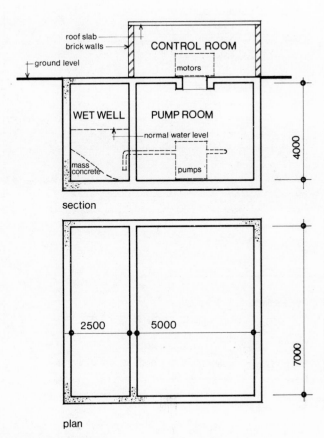

Figure 6.1 Layout of pumphouse

ref.	calculations EXAMPLE 6·1. SHEET 1.	output

DESIGN OF PUMPHOUSE.

Design the reinforced concrete underground
pumphouse shown in FIG. 6.1.

SOIL PROPERTIES.

      Granular soil : density = $18\,kN/m^3$
      Angle of repose   = $30°$
      Surcharge due to loaded
      vehicles on surrounding ground = $10\,kN/m^2$

Water in the wet well is normally at a level 2·0m.
above the floor, but the design should allow for
overflow conditions with the compartment full.
The pumphouse is dry.

DESIGN ASSUMPTIONS.

a) Loads.
   Soil pressure = $\dfrac{1-\sin \varnothing}{1+\sin \varnothing} \times 18 = \dfrac{1}{3} \times 18$

                  = $6\,kN/m^2$.

Although no water is said to be present in the
ground, the construction of the structure will
create conditions which will allow ground water
to collect.
A nominal allowance of 1 metre height of ground
water will be taken.
Assume density of water = $10\,kN/m^3$.

b) Design to BS. 5337.
Exposure class B for walls and floor.
Exposure class A. for roof slab to wet well.
Exposure class C. for roof to pumphouse.

ref.	calculations	output
	EXAMPLE 6·1. SHEET 2.	
BS 5337 TABLE 8.	c) <u>Materials</u>. Concrete grade 30 with a minimum cement content of 360 kg/m³ of finished concrete. (note this mix used for all concrete for simplicity). Reinforcement steel :- ribbed high yield bars grade 425.	CONCRETE GRADE 30
BS 5337 4·10.	d) <u>Cover</u> to outer layer of steel = 40mm.	COVER = 40.
	e) <u>Design</u>. Design all floor, wall and roof slab panels as continuous and 2-way spanning.	
	f) <u>Joints</u>. In view of the size of the structure, no movement joints are desirable as they are potential sources of leakage. The structure will therefore be designed as a monolithic structure, and construction joints will be shown on the drawings. No waterbar will be necessary at these joints, but the joint surface will be scabbled.	

ref.	calculations	output
	EXAMPLE 6·1. SHEET 3.	

<u>Loading Cases</u>

a) Wet Well Empty.

External soil and water pressures :-

Soil.	Ground Water	Surcharge.
6H.	10 (net 6.67).	$\frac{10}{3}$

Where ground water is present, the effective density
of the soil is reduced due to the buoyancy effect.
The effective extra load due to the presence of ground
water is therefore :—

$$10 \left(1 - \frac{1}{3}\right) H = 6.67 H.$$

b) Wet Well Full.

Internal pressure due to water in well.

10 H.

No allowance will be made for the passive resistance
of the external soil.

c). Partial Safety Factor for Loads = 1·6.

ref.	calculations EXAMPLE 6.1. SHEET 4.	output

<u>Thickness of Sections.</u>

For ease of construction of a wall 4.0m. high,
the minimum thickness should be 300mm. The
allowable ultimate shear strength of 30 grade concrete
with an assumed 0.5% of reinforcement is 0.55 N/mm².
The maximum ultimate shear force at the foot of the
walls due to the maximum external loading is :—

$$\frac{5}{8} \times 1.6 \left(\frac{1}{2} \times 6 \times 4^2 + \frac{1}{2} \times 6.67 + 3.33 \times 4 \right).$$
$$\text{soil.} \qquad \text{water} \qquad \text{surcharge.}$$

$$= 65 \text{ kN/m.}$$

The minimum effective depth of wall required for
shear is :-
$$d = \frac{65 \times 10^3}{0.55 \times 10^3} = 118 \text{ mm.}$$

Overall thickness = 118 + 10 + 40 = 168 mm.

Use walls 300 thick and floor 400 thick.

The roof to the wet well has to carry the surcharge
pressure of 10 kN/m² and is designed to
exposure class A.

Use a slab 250 thick.

Effective depth d = approx. 250 - 40 - 10.
$$= 200.$$
Span/effective depth = $\frac{2500}{200}$ = 12.5.

Satisfactory.

ref. column:

TABLE
3.5.1.

FACTOR $\frac{5}{8}$
IS FOR
APPROX
PROPPED
CANTILEVER
$\gamma f = 1.6$

CP.110.
3.3.8.1.

output column:

WALLS
h=300
FLOOR
h=400

ROOF
h=250.

ref.	calculations EXAMPLE 6·1. SHEET 5.	output

Calculation of Minimum Reinforcement.

Maximum length of continuous construction =
$$2500 + 5000 + 3 \times 300 = 8400 \, mm.$$

$T_1 = 30°C.$

$T_2 = 0°C$

$\alpha_e = 12.$

Net effective contraction strain
$$= 0.5 \times 12 \times 30 = 180 \, microstrain$$

Maximum allowable crack width $= 0.1mm$ (in roof slab).

$$\therefore S_{max} = \frac{0.1}{180 \times 10^6} = 926$$

Also $S_{max} = \frac{f_{ct}}{f_b} = \frac{\emptyset}{2\rho}$

Assume $\emptyset = 12 \, mm.$

$$S_{max} = \frac{2}{3} \times \frac{12}{2\rho} = \frac{4}{\rho}$$

$$\therefore \rho = \left(\frac{4}{926}\right) \times 100 = 0.43\%$$

Also.
$$\rho_{crit} = \frac{f_{ct}}{f_y} = \frac{1.3}{425} \times 100 = 0.31\%$$

$\therefore$ Minimum area of reinforcement
$$= 0.43\% \times 1000 \times 300 = 1290.$$

Use Y12 at 175 each face (1292).

Output column:
Y12 @ 175.
E.F.

ref.	calculations	output

EXAMPLE 6·1. SHEET 6.

Forces on Walls

```
        D         E
   ┌────────┬─────────┐
   │  WET   │ DRY.    │
 A │  WELL  │B        │ C
   │        │         │
   └────────┴─────────┘
        D         E.
```
PLAN.

Effective height of walls, centre/centre of
roof slab/floor slab = 4000 + 300 = 4300
The wall panels are loaded triangularly due to water
and soil pressures, and rectangularly due to surcharge
pressure.
It is convenient to replace the surcharge pressure
by the equivalent soil pressure.

4·3		4·3		4·85	

$$25·8 \quad + \quad 3·3 \quad = \quad 29·1.$$
$$\text{soil} \quad + \quad \text{surcharge} \quad = \quad \text{Equivalent.}$$

$$h \text{ equivalent} = h + \frac{3·3}{6·0} = 4·85.$$

Loading
Case 1.

External soil pressure
(including surcharge)   = 29·1.

Ground water   = 6·7

Case 2.

Internal water pressure = $10 \times 4·3 = 43·0.$

ref.	calculations EXAMPLE 6·1. SHEET 7	output

Wall A.

Case 1.

height of wall = 4300 ⎫ ratio l_x/l_z
length of wall = 7300 ⎭ = 1·7

M_V = vertical span
M_H = horizontal span
+M = tension on unloaded face.
–M = tension on loaded face.

APPENDIX B
CASE 2
FIG B1.

Soil – M_V = ·045 × 29·1 × 4·85 × 4·3 = 27·3
     + M_V = ·030 × 29·1 × 4·85 × 4·3 = 18·2
     – M_H = ·010 × 29·1 × 7·3² = 15·5
     + M_H = ·005 × 29·1 × 7·3² = 7·8

TREAT AS
CANTILEVER.

Ground water – M_V = $\frac{1}{2}$ × 6·7 × 1·0² = 3·4
Total.   – M_V = 27·3 + 3·4 = 30·7

Case 2.

Internal water:
– M_V = ·045 × 43 × 4·3² = 35·8
+ M_V = ·030 × 43 × 4·3² = 23·8
– M_H = ·010 × 43 × 7·3² = 22·9
+ M_H = ·005 × 43 × 7·3² = 11·5

Wall B.

As Case 2 Wall A on each face.

Wall C.

As Case 1 Wall A.

ref.	calculations $\qquad$ EXAMPLE 6·1. SHEET 8.	output
	Wall D	
	Case 1	
	height of wall $= 4300$ $\Big\}$ ratio. L_x/L_z.	
	length of wall $= 2800$ $= 0.65$	
	Symbols and conventions as Wall A.	
APP. B. FIG. B-1. CASE 2.	Soil $-M_V = \cdot01 \times 29 \cdot 1 \times 4 \cdot 85 \times 4 \cdot 3 = 6 \cdot 1.$	
	$+M_V = \cdot006 \times 29 \cdot 1 \times 4 \cdot 85 \times 4 \cdot 3 = 3 \cdot 7$	
	$-M_H = \cdot04 \times 29 \cdot 1 \times 2 \cdot 8^2 \qquad = 9 \cdot 1.$	
	$+M_H = \cdot025 \times 29 \cdot 1 \times 2 \cdot 8^2 \qquad = 5 \cdot 7$	
	Ground water (treat as cantilever)	
	$-M_V = \frac{1}{2} \times 6 \cdot 7 \times 1 \cdot 0^2 \qquad = 3 \cdot 4.$	
	Total $-M_V = 6 \cdot 1 + 3 \cdot 4 \qquad = 9 \cdot 5$	
	Case 2.	
	Internal water	
	$-M_V = \cdot01 \times 43 \times 4 \cdot 3^2 \qquad = 8 \cdot 0.$	
	$+M_V = \cdot006 \times 43 \times 4 \cdot 3^2 \qquad = 4 \cdot 8$	
	$-M_H = \cdot04 \times 43 \times 2 \cdot 8^2 \qquad = 13 \cdot 5$	
	$+M_H = \cdot025 \times 43 \times 2 \cdot 8^2 \qquad = 8 \cdot 4.$	
	Wall E.	
	height of wall $= 4300$ $\Big\}$ ratio $\frac{L}{H} = 1 \cdot 2.$	
	length of wall $= 5300$	
	Symbols and conventions as Wall A.	
APP B FIG. B-1. CASE 2.	Soil $-M_V = \cdot03 \times 29 \cdot 1 \times 4 \cdot 85 \times 4 \cdot 3 = 18 \cdot 2$	
	$+M_V = \cdot02 \times 29 \cdot 1 \times 4 \cdot 85 \times 4 \cdot 3 = 12 \cdot 1$	
	$-M_H = \cdot02 \times 29 \cdot 1 \times 5 \cdot 3^2 \qquad = 16 \cdot 3$	
	$+M_H = \cdot01 \times 29 \cdot 1 \times 5 \cdot 3^2 \qquad = 8 \cdot 2$	
	Cantilever : Water $-M_V = \frac{1}{2} \times 6 \cdot 7 \times 1 \cdot 0^2 = 3 \cdot 4.$	
	Total $-M_V = 18 \cdot 2 + 3 \cdot 4 = 21 \cdot 6.$	

ref.	calculations EXAMPLE 6·1. SHEET 9.	output

<u>Direct Forces</u>

All external loads cause compressive horizontal
forces in the walls which are resisted by the concrete
in compression. The value of the compressive stress is
low and may be ignored. The load due to the water
in the pumpwell causes tension in the walls which is
evaluated below.

The maximum tension is in Wall D. due to pressure on
Walls A and B.
Maximum water pressure = $10 \times 4.3 = 43$ kN/m².
Average water pressure over lowest 1 m. height of wall

$$= \frac{10\,(4.3 + 3.3)}{2}$$

$$= 38 \text{ kN/m}^2.$$

Total force = $38 \times 7.0\text{m} = 266$ kN/m height.
Force per metre height on each of walls A and B

$$= \frac{266}{2} = 133 \text{ kN}.$$

This calculation neglects the effect of the floor but
is conservative.

Area of steel required each face at $f_s = 200$

$$= \frac{133 \times 10^3}{2 \times 200} = 332 \text{ mm}^2.$$

ref.	calculations EXAMPLE 6·1. SHEET 10.	output

Floor Slab

The soil pressure under the floor slab is due to the
imposed weight of the structure.
The weight of the pumphouse super-structure (roof,
walls and floor slab) has been calculated separately
and amounts to 2200 kN.
Assuming a uniform distribution over the floor area,
the soil pressure is :-

$$\frac{2200}{5\cdot6 \times 7\cdot6} = 52 \text{ kN/m}^2.$$

(wet well area is neglected as load is largely over
pump well)

Floor slab spans 2 ways

$$L_y/L_x = \frac{7300}{5300} = 1\cdot4.$$

Assume simply supported and allow for fixing moments
later.

CP110
TABLE 13
CASE 9.

$+M_x = \cdot085 \times 52 \times 5\cdot3^2 = 124.$

$+M_y = \cdot056 \times 52 \times 5\cdot3^2 = 82$

Minimum fixing moments from walls (neglecting surcharge)

EX.6·1
(8).

Walls E : $M_F = -14\cdot1.$

Walls B & C: $M_F = -21\cdot5$

$Max +M_x = 124 - 14\cdot1 = 110 \text{ kNm}.$

$Max + M_y = 82 - 21\cdot5 = 60 \text{ kNm}.$

Mx   My.

ref.	calculations EXAMPLE 6·1. SHEET 11.	output

Reinforcement.

Wall thickness   = 300

Cover   =   40

Effective depth of inner layer of reinforcement

$$= 300 - 40 - 12$$

$$= 248.$$

Minimum reinforcement calculated on sheet 6.1 (5) is Y12 at 175 each face.

Referring to appendix A table A 2.3, there is no entry for Y12 at 175 at the top of the table where f_s increases. Taking a desirable maximum service stress of $f_s = 200$ N/mm^2, the crack width will either be less than 0.2 mm or the section may be uncracked. The section may therefore be designed for strength using a cracked section and service stress of $f_s = 200$.

Using formula (4) section 3.6.1.

$$P = \frac{A_s}{bd} = \frac{646}{10^3 \times 242} = 0.00267$$

$$x_1 = 15 P \left(\sqrt{1 + \frac{2}{\alpha_e P}} - 1 \right)$$

$$z_1 = \left(1 - \frac{x_1}{2} \right) = 0.877$$

$$z = z_1 d = 0.877 \times 248 = 217$$

$$M = A_s f_s z$$
$$= 646 \times 200 \times 217 \times 10^{-6}$$
$$= 28 \text{ kNm/m}.$$

ref.	calculations	output
	EXAMPLE 6·1. SHEET 12.	

Wall A.

Ex 6·1.
(11).

Horizontal steel, all moments are less than
28 kNm/m therefore, for bending, use
Y12 at 175 A_s = 646. Add for direct tension
A_s required = 646 + 332 = 978.
Use   Y16 at 200 E.F

WALL A.
Y16 at
200 EF.

Vertical steel
M = 30·7 (this is maximum value : use same
                          steel each face).

Table A 2.3.
Try Y16 at 175 E.F.
Allowable M = $\frac{1}{2} \times \left(\frac{200}{210} \times 60 + \frac{200}{230} \times 50 \right)$
                      = 50
                              Satisfactory

WALL A.
Y16 at
175 EF.

Wall B.   As Wall A.

Wall C   As Wall A.

Wall D.   Y12 at 175 EF, EW.

WALL D.
Y12 at
175 EW,EF

Wall E   Y12 at 175 EF, EW.
               (Note EF = each face, EW = each way)

WALL E.
Y12 at
175 EW,EF.

Floor thickness = 400
Cover   = 40.
Effective depth of inner layer = 400 − 40 − 16 − 8 = 336.
Negative moments as wall steel

$+M_x$ = 110
Table A 2.5   Use Y20 at 150 TF   (2090)

FLOOR
Y20 at
150. TF.

ref.	calculations EXAMPLE 6·1. SHEET 13.	output
	$+ M_y = 60$ <u>Use Y16 at 150 TF</u> (1340).	Y16 at 150 TF.

FIG. 3.11.	By inspection, crack control is not critical.
	
	Using $f_s = 200$ $d = 400-40-20-8$
	                       $= 332.$
	$\dfrac{M \times 10^2 \times 10^6}{f_s b d^2} = 0.27.$
	
	$\therefore P = 0.3\%.$
	Minimum steel in floor.
	    $S_{max} = \dfrac{0.2}{180 \times 10^6} = 1850\,mm.$
	
	    $\emptyset = 20.$
	    $S_{max} = \dfrac{2}{3} \times \dfrac{20}{2P} = \dfrac{20}{3}P = 1850$
	
	    $\therefore P = 0.36\%$
	     $P_{crit} = 0.31\%$
	
	    $\therefore A_s = 0.36 \times 10 \times 332 = 1195$
	Use <u>Y12 at 175 EF</u> (1292)
		Y12 at 175 EF.

ref.	calculations	output
	EXAMPLE 6·1. SHEET 14.	

<u>REINFORCEMENT DETAILS</u>

<u>WALL A</u>

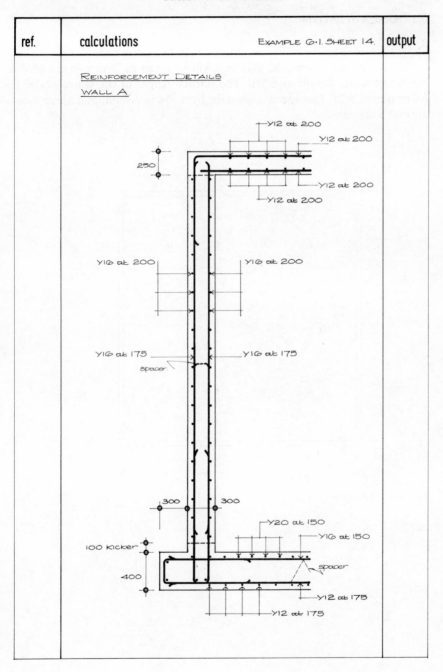

6.2 Design of reservoir

Design a roofed reservoir to contain 4 000 000 litres of water.

Due to site conditions, the plan size will be taken as 26 m × 26 m and the maximum water height as 6.5 m. The normal height of the stored water is 6.0 m (figure 6.2). The site is underlain by a granular soil, and there is no ground water present.

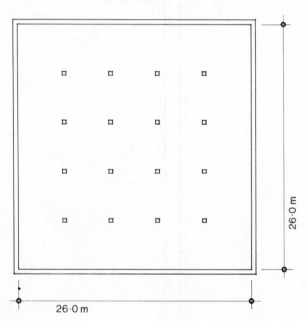

Figure 6.2   Layout of reservoir

ref.	calculations	output
	EXAMPLE 6.2 SHEET 1	

DESIGN OF RESERVOIR.

Layout of structure.

The size of the structure is less than the length which would require an expansion joint (say 70m).

Contraction joints will be provided at approx 5.0m centres in the wall and floor panels.

The roof will be designed as a flat slab spanning between columns at approx 5.0m centres as shown in fig. 6.2.

Provide a sliding joint between the top of the wall and the roof slab.

The cantilever wall is designed for water pressure in detail on pages ex. 6.2/2 to 4. The design against soil pressure is prepared from the design charts.

Exposure Conditions.

Wall inner face   Class B.

Wall outer face   Class C

Floor   Class B

Roof   Class A.

_____

Materials

Concrete   grade 25.

Steel   grade 425/460.

Partial Factors of Safety.

Materials.

        Concrete   1.5.

        Steel   1.15

Loads.

        Water.   1.6

        Soil   1.6.

ref.	calculations EXAMPLE 6.2 SHEET 2.	output

Limit State Design

Cantilever Wall.

Design a cantilever wall to withstand
water pressure.

Height $H = 6.5$ m.

Water $W_g = 10 kN/m^3$

Partial safety factors :—

Ultimate limit state — $\delta_F = 1.6$

Service limit state — $\delta_F = 1.0$.

TABLE 3.1
TABLE A2.5

From design charts the estimated root thickness
$h = 800$ mm and the tension reinforcement is
Y25 at 150   (3270)

Effective depth $d = 800 - 40 - 12.5 = 735$.

output: $h = 800$. Y25 at 150.

Ultimate limit state.

Applied moment $M = \frac{1}{6} \times (1.6 \times 10) \times 6.5^3$

$\qquad\qquad = 732$ kNm/m.

The ultimate moment of resistance based on the
concrete is :—

SECTION 3.5(4)

$M_U = 0.157 f_{cu} b d^2$

$\qquad = 0.157 \times 25 \times 10^3 \times 735^2 \times 10^{-6}$

$\qquad = 2120$ kNm/m   Satisfactory.

The ultimate moment of resistance based on the
steel is given by

SECTION 3.5(3).

$M_U = (0.87 f_y) A_s Z$

where the lever arm $Z = d - 0.45 x$

$\qquad Z = 687$

$\qquad M_U = 830.$

output: ULTIMATE MOMENT SATISFACTORY

The section is therefore satisfactory as the applied
ultimate moment is less than the moment of resistance
of the section.

ref.	calculations EXAMPLE 6·2 SHEET 3.	output
	Ultimate shear force at root of wall. $$V = \tfrac{1}{2} \times (1.6 \times 10) \times 6.5^2$$ $$= 338 \text{ kN/m}.$$ $\therefore$ shear stress $$v = \frac{338 \times 10^3}{1000 \times 735} = 0.46.$$ $$\frac{100 A_s}{bd} = \frac{100 \times 3270}{1000 \times 735} = 0.44$$	
CP 110 TABLE 5.	$\therefore v_c = 0.464$ (by interpolation) This is satisfactory and no shear steel is necessary.	ULTIMATE SHEAR SATISFACTORY
	<u>Limit State of Cracking</u>. Service moment $M_s = \tfrac{1}{6} \times 10 \times 6.5^3$ $\qquad\qquad = 458 \text{ kNm}.$ Depth of neutral axis (elastic no-tension theory) $$x_1 = \alpha_e P \left(\sqrt{1 + \tfrac{2}{\alpha_e P}} - 1 \right)$$ modular ratio $\alpha_e = 15$. $$\frac{A_s}{bd} = P = \frac{3270}{1000 \times 735} = 0.00445$$ $\therefore \alpha_e P = 15 \times 0.00445 = 0.0667$. ($\alpha_e = 15$ is sufficiently accurate for grades 25 and 30 concrete). $\therefore \qquad x = 735 \times 0.30 = 224$	
	Moment of resistance of section $$M_r = A_s f_s \times (d - \tfrac{x}{3})$$ $\therefore \quad 458 = 3270 \times f_s \times (735 - \tfrac{1}{3} \times 224)$ $\therefore \qquad f_s = \underline{212}.$	$f_s = 212.$

ref.	calculations EXAMPLE 6·2 SHEET 4.	output
BS 5337 App. C. PARA. 1.	Check steel and concrete service stresses. $$f_s = 212 \qquad 0.8 f_y = 0.8 \times 425 = 304.$$ $$f_{cb} = \frac{212}{15} \times \frac{224}{(735-224)} = 6.9$$ $$0.45 f_{cu} = 0.45 \times 25 = 11.5.$$ Satisfactory. Elastic strain at surface. $$\varepsilon_1 = \frac{h-x}{d-x} \times \frac{f_s}{E_s}$$	
CP 110 2.4.2.4.	$$= \frac{800-224}{735-224} \times \frac{212}{200} \times 10^{-3}.$$ $$= 1.195 \times 10^{-3}$$ $$\varepsilon_m = 1.195 \times 10^{-3} - \frac{0.7 \times 1000 \times 800}{3270 \times 212} \times 10^{-3}.$$ $$= 0.387 \times 10^{-3}$$	
BS 5337 App. C (1).	Crack width $w = \dfrac{4.5\, a_{cr}\, \varepsilon_m.}{1 + 2.5 \left(\dfrac{a_{cr} - C_{min.}}{h-x} \right)}$ bar spacing $S = 150$ $C_{min.} = 52$ $C_a = 52 + 12.5 = 65.$ $C_{min} = 40 + 12 = 52$ bar size $\phi = 25.$ $a_{cr} = \sqrt{\left(\frac{S}{2}\right)^2 + C_a^2} - \frac{\phi}{2} = 86.4.$ ∴ surface crack width $$w = \frac{4.5 \times 86.4 \times 0.367 \times 10^{-3}}{1 + 2.5 \left(\frac{86.4 - 52}{800 - 224} \right).}$$ $$= 0.131$$ Allowable $w = 0.2$ mm. Satisfactory.	$w = 0.131.$

ref.	calculations EXAMPLE 6.2 SHEET 5.	output

<u>Cantilever Wall</u>   (reservoir empty, soil pressure only).

soil density = 18 kN/m³

angle of repose = 30°

∴ soil pressure = 6h.

surcharge due to soil and imposed

load = 0.6 × 18 + 5 = 15.8 kN/m²

6.5

6H.   15.8

Total service moment at base of wall

$$= \frac{1}{6} \times 6 \times 6.5^3 + \frac{1}{2} \times 15.8 \times 6.5^2$$

= 608 kNm.

Ultimate moment

= 1.6 × 608

= 973 kNm.

From previous calculation wall thickness h = 800

cover   = 40 + 20 = 60.

effective depth = 800 − 40 − 12 − 12.5

= 735.

Class C exposure

$$M_u = 0.157 \times 25 \times 10^3 \times 735^2 \times 10^{-6}$$

= 2120 kNm.

Depth of neutral axis

$$= x = \frac{735\left(1 - \sqrt{1 - 0.70\frac{M}{M_u}}\right)}{0.9}$$

= 144

Lever arm = 735 − 0.45 × 144

= 670

$$A_s = \left(\frac{M}{0.87 \times 425 \times 670}\right) = \frac{973 \times 10^6}{(0.87 \times 425) \times 670}$$

= 3927

<u>Use Y25 at 125</u>   (3930).

Y25 at
125.

ref.	calculations	output
	EXAMPLE 6·2 SHEET 6	

ref.	calculations	output
	<u>Cantilever Wall</u> — distribution steel.	
	Use limit state design.	
	Provide movement joints at 5·2 m. centres.	
	Critical steel ratio $P_{crit} = \dfrac{f_{ct}}{f_y}$	
	For grade 25 concrete the direct tensile strength at 3 days $f_{ct} = 1·15 \text{ N/mm}^2$	
	For grade 425 steel $f_y = 425 \text{ N/mm}^2$.	
	$P_{crit} = \dfrac{1·15}{425} = 0·27\%$	
	For close joint spacings, minimum steel ratio	
	$= \frac{2}{3} P_{crit} = 0·18\%$	$P_{min} = 0·18\%$
	For class B exposure, maximum permissible design crack width = 0·2 mm.	
	Also $W_{max} = S_{max} \times \dfrac{\alpha}{2} T$	
	$\alpha = 12$ microstrain/°C.	
	T is assumed to be 40°C (for a wall over 500 thick)	
BS 5337 B3.	For walls over 500 thick, only 500 need be considered in calculating the distribution reinforcement.	
	At base level $P = 0·22\%$ ($\varnothing = 16$).	
	$A_s = 0·22\% \times 800 \times 1000 = 1760 \text{ mm}^2/\text{m}.$	Y16 at 200 EF.
	<u>Use Y16 at 200 each face</u> (2020)	
	As the reinforcement remains the same until the thickness reduces below 500, the same distribution reinforcement will be used for the full height of the wall.	

ref.	calculations EXAMPLE 6·2 SHEET 7.	output
	Roof Slab.	

<u>Roof Slab.</u>

Design as flat slab with continuous construction,
(ie. no movement joints).
Provide movement joint at junction of slab and walls.

Loads.

Imposed due to light construction traffic.	5.0	
600 soil cover	10·8	
	15·8	kN/m².

With a careful construction specification, the
imposed load can be taken as either not present,
or present on all spans simultaneously.
The critical span is adjacent to the external walls.
Due to the assumed incidence of loading, there will
be only a small transfer of moment to the columns
which will be ignored in the design of the slabs.

Assume slab thickness = 450.
Dead load = 24 × 0·450 = 10·8.
Imposed load =   15·8
                                                  26·6 kN/m².

Maximum +M for each span

CP 110
TABLE 4.

$$+M = \frac{1}{11} \times (26·6 \times 5·2) \times 5·2^2$$

$$= 340 \, kNm / full \, bay \, width.$$

ref.	calculations EXAMPLE 6·2 SHEET 8.	output

Divide slab into column strips and middle strips, each 2·6 m. wide.

CP110
TABLE 17.
75% on
COLUMN
STRIP.

+ Moment/metre width on column.

$$\text{Strip} = 340 \times \frac{0·75}{2·6}$$

$$= 98 \text{ kNm/m.}$$

Design for class A exposure.

Table A 1.6.

$h = 450$ Use Y20 at 200 (1570) *.

$M_s = 107·6$ $(f_s = 200)$.

Cover to lowest layer = 40.

Cover to second layer = 60

∴. Results from table are satisfactory.

(* But see below for distribution steel calculation.)

Roof slab — minimum reinforcement.

As before $P_{crit} = \frac{1·15}{425} = 0·27\%$.

Assume :—

$$T = 30°C$$

$$\alpha = 12 \times 10^{-6}$$

$$\omega = 0·1 \text{ mm.}$$

$$S_{max} = \frac{2}{3} \times \frac{\phi}{2P}$$

also $S_{max} = \dfrac{\omega}{\epsilon} = \dfrac{0·1}{\frac{1}{2} \times 12 \times 10^{-6} \times 30} = 556$

∴ $\dfrac{\phi}{P} = 556 \times 3 = 1666.$

∴ P is given by

ϕ	$P\%$	A_s (each face).
12	0·72	1620
16	0·96	2160
20	1·20	2700.

Y16 at 100 provides an area of 2010 — In view of the variations in the practical situation this is satisfactory.

Use this arrangement of steel both faces and in both directions to provide both main and shrinkage steel.

Y16 at 100.
EW. EF.

ref.	calculations	output

calculations — EXAMPLE 6·2 SHEET 9.

CP110 TABLE 4.

<u>Shear at Column Head.</u>

Maximum column load

$$= 26 \cdot 6 \times 5 \cdot 2 \times 5 \cdot 2 \times 1 \cdot 15$$

$$= 827 \text{ kN}.$$

Assume that $\gamma_f = 1 \cdot 4$ for all applied loading, as imposed load (after construction) will be unusual.

Ultimate shear force

$$= 1 \cdot 4 \times 827$$

$$= 1158 \text{ kN}.$$

CP110 3.6.2.

Allow for unequal distribution

Ultimate shear force

$$= 1 \cdot 25 \times 1158$$

$$= 1448 \text{ kN}.$$

CP110 3.4.5.2.

Assume column size $= 400 \times 400$.

Slab thickness $h = 450$.

Critical perimeter

$$= 4 \times 400 + 9 \cdot 4 \times 450$$

$$= 5830.$$

Steel ratio $P = \dfrac{100 As}{bd} = \dfrac{100 \times 2010}{1000 \times 390}$

$$= 0 \cdot 52.$$

CP110 TABLE 5.

Concrete grade 25

Permissible shear stress $= v_c = 0 \cdot 51$.

∴ permissible shear force

$$= 0 \cdot 51 \times 5830 \times 390 \times 10^{-3}$$

$$= 1160 \text{ kN}.$$

Satisfactory.

<u>Columns.</u>

Maximum ultimate column load

$$= 827 \times 1 \cdot 4$$

$$= 1160 \text{ kN}.$$

Column height $= 6 \cdot 5 \text{ m}$.

Effective height $L_e = 1 \cdot 5 \times 6 \cdot 5$ (unbraced)

$$= 9 \cdot 75 \text{ m}.$$

output: NO SHEAR STEEL.

ref.	calculations EXAMPLE 6·2 SHEET 10.	output
	Assume column size 400×400.	
	Slenderness ratio = $\dfrac{Le}{h}$	
CP110 3.5.1.2 3.5.7.1.	$= \dfrac{9750}{400} = 24.$	
	∴ column is slender	
	$M_t = 0.05 N_h + \dfrac{Nh}{1750}\left(\dfrac{Le}{h}\right)^2\left(1 - 0.0035\dfrac{Le}{h}\right).$	
CP110 FORMULA 38.	$= 23.2 + 140 = 163.2.$	
	$K = 0.5$	
	∴ effective $M_t = 0.5 \times 163.2 = 82$ kNm.	
	$\dfrac{N}{bh} = \dfrac{1160 \times 10^3}{400^2} = 7.25.$	
	$\dfrac{M}{bh^2} = \dfrac{82 \times 10^6}{400^3} = 1.28.$	
CP110 PART 2. CHART 76.	$\dfrac{100 Asc}{bh} = 0.1.$	
	Minimum steel = 1%.	
	$A_{sc} = 400^2 \times 0.01 = 1600$	
	<u>Use 4 - Y25</u> (1960).	4-Y25 R8 at 300.
	<u>Links R8 at 300.</u>	
	<u>Column Bases</u>	
	Service load = 827	
	S.W. column = 25	
	S.W. base * = 30 * extra over soil.	
	$\overline{882}$ kN.	
	Use base 2250 × 2250 × 1000.	
	Moment due to slender column = 82.	
	Soil pressure under base	
	$= \dfrac{882}{2.25^2} \pm \dfrac{82 \times 6}{2.25^3}$	
	$= 174 \pm 43$	MAX. SOIL PRESSURE = 217 kN/m².
	$= 217$ or 131 (kN/m²) Satisfactory.	
	Note : total pressure	
	$= 217 + $ water pressure	
	$= 217 + 10 \times 6.5$	
	$= 282$ kN/m².	

ref.	calculations EXAMPLE 6·2 SHEET 11.	output

Wall Footings

Water pressure = 10h.

Soil pressure = 6h.

Surcharge = 15·8 kN/m².

Roof slab load = 26·6 × 5·2 = 138 kN/m.

D.L. only = 10·8 × 5·2 = 56 kN/m.

Case 1

Tank full — no soil.

Take moments about A.

Restoring forces.

  Water $10 \times 6·5 \times 2·2 = 143 \times 3·9 = 558$

  Wall $24 \times 6·5 \times 0·6 = 94 \times 2·4 = 225$

  Base $24 \times 5·0 \times 0·8 = 96 \times 2·5 = 240$

  Roof $56 \times 2·4 = 134$

                           $\overline{389}$ $\overline{1157}$

Overturning moment

     $= \frac{1}{6} \times 10 \times 6·5^3$ $= \underline{457}$

                                $\overline{700}$

Factor of safety against overturning

     $= \frac{1157}{457}$

     $= 2·53$ Satisfactory

                 $(> 2·0)$.

ref.	calculations	output
	EXAMPLE 6·2 SHEET 12.	

Case 2.

Tank empty — soil + surcharge

Take moments about B.

Restoring forces.

Soil	$18 \times 6.5 \times 2.0$	$= 234 \times 4.0$	$= 936.$	
Surcharge	15.8×2.0	$= 32 \times 4.0$	$= 126.$	
Wall	$24 \times 6.5 \times 0.6$	$= 94 \times 2.6$	$= 243$	
Base	$24 \times 5.0 \times 0.8$	$= 96 \times 2.5$	$= 240.$	
Roof		$= 138 \times 2.6$	$= 359$	
		$\overline{594}$	$\overline{1904.}$	

Overturning moment

$$= \tfrac{1}{6} \times 6 \times 6.5^3 + \tfrac{1}{2} \times 15.8 \times 6.5^2 = \quad 608$$
$$\text{(soil)} \qquad \text{(surcharge).} \quad \overline{1296}$$

Factor of safety against overturning

$$= \frac{1904}{608} = 3.1 \quad \text{Satisfactory.}$$
$$(> 2.0).$$

Case 3.

Tank full + soil + surcharge.

Take moments about B.

Restoring forces.

As case 2		594	1904
Water	$143 \times 1.1 =$	157	
		$\overline{737}$	$\overline{2061.}$

Net overturning moment about B

$$= 608 - 457 \qquad = \quad 151$$
$$\overline{1910.}$$

By inspection — not critical.

ref.	calculations EXAMPLE 6·2 SHEET 13	output

Soil pressure under base:

Calculate moment about centre line of base.

Case 1.

$$M_C = 389 \left(\frac{700}{389} - 2·5 \right) = 273.$$

Soil pressure

$$f = \frac{389}{5·0} \pm \frac{273 \times 6}{5·0^2}$$

$$= 78 \pm 66$$

$$= 144 \text{ or } 12.$$

Case 2.

$$M_C = 594 \left(\frac{1296}{594} - 2·5 \right) = 189.$$

Soil pressure

$$f = \frac{594}{5·0} \pm \frac{189 \times 6}{5·0^2}$$

$$= 119 \pm 45$$

$$= 164 \text{ or } 74.$$

Case 3

$$M_C = 737 \left(\frac{1910}{237} - 2·5 \right) = 68.$$

Soil pressure

$$f = \frac{737}{5·0} \pm \frac{68 \times 6}{5·0^2}$$

$$= 148 \pm 16$$

$$= 164 \text{ or } 132.$$

Maximum soil pressure = 164.

Satisfactory.

SOIL PRESSURE = 164.

ref.	calculations EXAMPLE 6.2 SHEET 14.	output

Footing Reinforcement

Case 1.

Moment at root of toe

$$= (65.0 + 19.2) \times 2.2^2 \times 0.5 -$$
$$0.5 \times 70 \times 2.2^2 \times \tfrac{1}{3} - 0.5 \times 12 \times 2.2^2 \times \tfrac{2}{3}$$

$$= 128 \text{ kNm/m}.$$

Moment at root of heel.

$$= 0.5 \times 144 \times 2.0^2 \times \tfrac{2}{3} + 0.5 \times 91 \times 2.0^2 \times \tfrac{1}{3} -$$
$$0.5 \times 19.2 \times 2.0^2$$

$$= 215 \text{ kNm/m}.$$

Case 2.

Moment at root of toe

$$= 0.5 \times 124 \times 2.2^2 \times \tfrac{1}{3} + 0.5 \times 104 \times 2.2^2 \times \tfrac{2}{3} -$$
$$0.5 \times 19.2 \times 2.2^2$$

$$= 318 \text{ kNm/m}.$$

ref.	calculations EXAMPLE 6.2 SHEET 15.	output
	Moment at root of heel $$= 0.5 \times (15.8 + 117 + 19.2) \times 2.0^2 -$$ $$0.5 \times 74 \times 2.0^2 \times \frac{2}{3} - 0.5 \times 110 \times 2.0^2 \times \frac{1}{3}$$ $$= 132 \text{ kNm/m}.$$ Case 3. By inspection, not critical. Heel reinforcement Exposure class C. Top reinforcement $$M = 132$$ $$M_U = 132 \times 1.6 = 211.$$ $$h = 800$$ $$d = 735$$ $$x_1 = \frac{\left(1 - \sqrt{1 - 0.7 \times \frac{211}{2120}}\right)}{0.9}$$ $$= 0.039$$ $$z_1 = 1 - 0.018 = 0.98$$ but maximum value = 0.95 $$\therefore A_{st} = \frac{211 \times 10^6}{0.95 \times 735 \times 0.87 \times 425} = 817.$$ Use <u>Y16 at 125 top</u> (1610) Bottom reinforcement $$M = 215$$ $$M_U = 215 \times 1.6 = 344$$ $$x_1 = 0.065$$ $$z_1 = 0.95$$ $$A_{st} = \frac{344 \times 10^6}{0.95 \times 735 \times 0.87 \times 425} = 1332$$ Use Y16 at 125 bottom (1610)	Y16 at 125. Y16 at 125.

ref.	calculations EXAMPLE 6.2 SHEET 16	output
TABLE A2.5. FIG. 3.11.	Toe reinforcement. Exposure class B Top reinforcement. $\qquad M = 128.$ $\qquad$ Cracking not significant Use $f_s = 200.$ $Factor = \dfrac{128 \times 10^2 \times 10^6}{200 \times 10^3 \times 735^2} = 0.12$ $P = 0.15\%$ $A_s = 0.15\% \times 10^3 \times 735 = 1103$ Use Y16 at 125 (1610) Bottom reinforcement. Exposure class C. $\qquad M = 318$ $\qquad M_u = 318 \times 1.6 = 509$ $x_1 = \dfrac{\left(1 - \sqrt{1 - 0.7 \times \dfrac{509}{2120}}\ \right)}{0.9} = 0.1.$ $z_1 = 1 - 0.45 x_1 = 0.95.$ $A_s = \dfrac{509 \times 10^6}{0.95 \times 735 \times 0.87 \times 425} = 1972$ Use Y25 at 125 to suit wall reinforcement. (3930)	Y16 at 125. Y25 at 125.

ref.	calculations EXAMPLE 6.2 SHEET 17	output

Reservoir Floor Slab.

Floor is divided into 5.2m. panels by movement joints.
Assuming uniform ground conditions, the floor slab is uniformly loaded and has no transverse bending stresses.

To provide a reasonable thickness of concrete, and allowing for possible construction tolerances, use a slab 200mm. thick.

$$P_{crit} = \frac{f_{ct}}{f_y} = \frac{1.15}{425} = 0.27\%$$

BS 5337
B.1.

Provide top reinforcement only based on a surface zone 100mm deep (ie. one half of slab depth).

$$A_s = 0.27\% \times 100 \times 1000$$

$$= 270 \, mm^2/m.$$

Use a welded mesh fabric with 10mm wires in each direction at 200mm spacings (393).

MESH 10
AT 200.

Bottom zone does not require reinforcement.

ref.	calculations EXAMPLE 6·2 SHEET 18.	output
	REINFORCEMENT DETAIL. 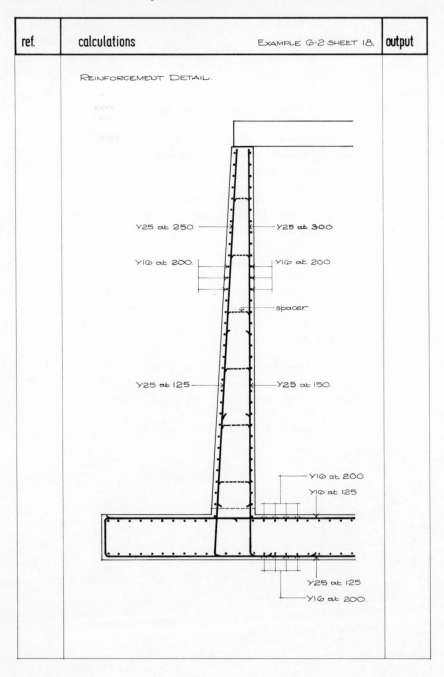	

6.3   Design of a circular prestressed concrete tank

Design a circular prestressed concrete tank to contain water for fire-fighting purposes (figure 6.3).

Tank diameter = 20.0 m
Height of water =   7.5 m

The tank is to be constructed entirely above ground level.

This example is intended to show the application of prestressed concrete design to liquid-retaining structures.

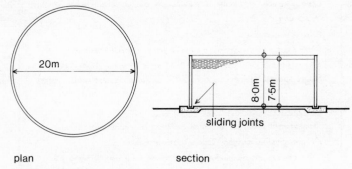

plan   section

Figure 6.3   Layout of circular pre-stressed concrete circular tank

ref.	calculations EXAMPLE 6·3 SHEET 1	output

DESIGN OF PRESTRESSED CIRCULAR TANK.

    Internal diameter of tank = 20·0m.

    Maximum depth of water   =   7·5m

    Allow freeboard of 0·5m.

    Tank is constructed above ground.

    Provide a sliding joint at the foot of the wall (Fig 6·3)

Materials

For prestressed concrete construction a high strength

concrete is required.

BS 5337
TABLE 8.

Use grade 40 with a minimum cement content of

300 kg/m³.

Reinforcement

Use high strength, low relaxation prestressing

strands to BS 5896:1980, and grade 425/460

high yield deformed reinforcement to BS 4461:1978.

Exposure Conditions.

BS 5337
4·9

The tank will be full of water for most of its life.

∴ exposure class B is appropriate.

BS 5337
10·1.

The basic requirement is to ensure that there is a

circumferential compression in the concrete of

1·0 N/mm² when the tank is full of water.

The prestressing cables will be placed outside the

walls and protected with sprayed concrete.

Concrete cover to normal reinforcement = 40.

Wall Thickness

To enable the concrete to be placed with 4 layers

of normal reinforcement and to prevent local

percolation, use a wall thickness of 225.

Output column:

40 GRADE
CONCRETE.

7 WIRE
STANDARD
STRAND

GRADE
425/460
STEEL.

COVER
= 40.

h =
225.

ref.	calculations EXAMPLE 6·3 SHEET 2.	output

PLAN.   WATER PRESSURE.

Maximum water pressure   = 10 × 7·5

                            = 75 kN/m²

Maximum circumferential tension due to water pressure

$$T = \frac{75 \times 20}{2} = 750 \text{ kN/m.}$$

BS 5337
10·1.

The extra force required to provide a net circumferential compressive stress in the concrete of 1 N/mm² is

$$T_1 = \text{Area} \times \text{stress}$$
$$= 225 \times 1000 \times 1·0 \times 10^{-3}$$
$$= 225 \text{ kN/m.}$$

Total force required

$$= T + T_1$$
$$= 750 + 225$$
$$= 975 \text{ kN/m.}$$

At top of tank water pressure = zero and T = Q.
Total force required is

$$T_1 = 225 \text{ kN/m.}$$

ref.	calculations EXAMPLE 6.3 SHEET 3	output

225

225
290   280 spacing
407   200 spacing
542   150 spacing
678   120 spacing
814   100 spacing
975   80 spacing

975

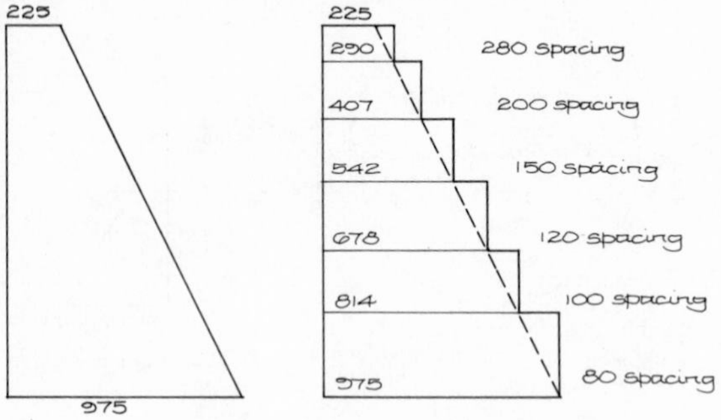

theoretical minimum
circumferential force
required.

actual forces provided by
tendons at varying spacings
up wall height. (Details
are calculated later on page
6.3 (8.)

BS 5896

Prestressing Cables.
Use 12.5 mm 7 wire low relaxation strand.

$$Area = 93 \, mm^2$$

Characteristic load = 164 kN.

$$Characteristic \ strength = \frac{164 \times 10^3}{93}$$

$$= 1770 \, N/mm^2$$

Initial jacking force = 0.7×164

= 115 kN.

Initial stress = 0.7×1770

= 1240 N/mm^2

output column:

$P_0 = 115.$

ref.	calculations EXAMPLE 6·3 SHEET 4	output

Stressing sequence.

$$x = \frac{\pi \times 20 \cdot 45}{8}$$

$$= 8 \cdot 0 \, m$$

x = 8·0m.

The prestressing strands are anchored on concrete projections (pillasters) at the quarter points of the circumference.

Each cable extends half-way round the circumference and is stressed from both ends. Both cables at each level are stressed simultaneously and four jacks are required.

The jacking points for alternate cables (in elevation) are at A and then B etc.

The maximum friction loss is half-way between jacking points for each cable and by averaging for two consecutive cables AA and BB, the point of average maximum loss will be at a point defined by $x = 8 \cdot 0m$.

ref.	calculations EXAMPLE 6·3 SHEET 5.	output

Loss of prestressing force due to friction.

  Assume constant for friction due to irregularities in ducts :-

$$k = 33 \times 10^{-4} /m$$

Coefficient of friction between tendon and duct.

$$\mu = 0·30.$$

Radius of curvature of tendons

$$R = \frac{20·45}{2}$$

$$= 10·2\,m.$$

Point of average maximum loss of stress

Factor $\left(kx + \frac{\mu x}{R} \right)$ $x = 8·0\,m.$

$$= 8·0 \left(33 \times 10^{-4} + \frac{0·3}{10·2} \right)$$

$$= 0·262$$

If the initial prestressing force $= P_0$, the force after friction losses at x is

$$P_x = P_0 \times e^{-0·262}$$

$$= 0·77 P_0.$$

$\therefore$ $P_x = 0·77 \times 115$

$$= 88·5\,kN.$$

and stress in strand after friction losses

$$= 0·77 \times 1240$$

$$= 955\,N/mm^2.$$

ref.	calculations EXAMPLE 6.3 SHEET 6.	output
	<u>Prestressing Losses</u>	
	1). Loss due to creep of low relaxation strand = 2% (from manufacturers' catalogue).	2·0
	2). Loss due to elastic contraction of concrete.	
CP 110 BS.5896	Modulus of elasticity of concrete = 31 kN/m^2 Modulus of elasticity of steel = 195 kN/m^2 Modular ratio $\alpha_e = \dfrac{195}{31}$ = 6.3	
	Maximum elastic stress in concrete with tank empty after losses $$= \frac{975 \times 10^3}{225 \times 10^3} = 4\cdot3 \text{ N/mm}^2$$	
	Assuming 10% total losses, concrete stress at transfer $$= \frac{4\cdot3}{0\cdot9} = 4\cdot8 \text{ N/mm}^2.$$	
	$\therefore$ elastic strain in concrete $$= \frac{4\cdot8}{31}$$ which is equal to loss of strain in steel $\therefore$ loss of stress in steel $$= \frac{4\cdot8 \times 195}{31}$$ or $4\cdot8 \times 6\cdot3 = 30\cdot2 \text{ N/mm}^2.$	
	As the tank will be post-tensioned, the final strands will be tensioned after nearly all the elastic shortening in the concrete has taken place, therefore the average loss may be taken as half the value calculated above. ie loss $= \frac{1}{2} \times 30\cdot2 = 15\cdot1 \text{ N/mm}^2$ Initial stress in strand = 955 N/mm^2 (after friction losses)	
	$\therefore$ % loss $= \dfrac{15\cdot1 \times 100}{955} = 1\cdot6\%.$	1·6%

ref.	calculations EXAMPLE 6·3 SHEET 7.	output

3). Loss due to shrinkage of concrete.

CP110
TABLE 41.

Shrinkage strain in concrete = 200×10^{-6}

∴ loss of strain in strands   = 200×10^{-6}

∴ loss of stress in strands   =

$$(strain) \times \alpha_{es} = 200 \times 10^{-6} \times 195 \times 10^{3}$$
$$= 39 \, N/mm^{2}.$$

$$\% \, loss = \frac{39 \times 100}{955} = 4.1\%$$

4.1%.

4). Loss of pressure due to creep of concrete.

Stress in concrete at transfer = $4.8 \, N/mm^{2}$.

Proportion of cube strength   = $\dfrac{4.8}{40}$
$$= 0.12.$$

As this is less than $\frac{1}{3}$, the creep values need not be increased.

CP110
4.8.2.5.

Creep strain = 36×10^{-6}.

Loss of stress in strand =

$$(strain) \times \alpha_{es} = 36 \times 10^{-6} \times 195 \times 10^{-3}$$
$$= 7.0.$$

$$\% \, loss = 7.0 \times \frac{100}{955} = 0.7\%$$

0.7%

Total losses = $(2.0 + 1.6 + 4.1 + 0.7)\%$
$$= 8.4\%.$$

TOTAL
LOSSES
8.4%.

This is less than the value of 10% which was assumed on page 6 and is satisfactory.

Stress in strand after friction losses = $955 \, N/mm^{2}$.

∴ final effective stress

$$= 955 \left(\frac{100 - 8.4}{100}\right) = 875 \, N/mm^{2}.$$

and force/tendon

$$= 875 \times 93 \times 10^{-3} = 81.4 \, kN.$$
$$(area)$$

EFFECTIVE
TENDON
FORCE
81.4 kN.

ref.	calculations	output
	EXAMPLE 6·3 SHEET 8.	

Number of tendons per metre at bottom of tank

$$= \frac{\text{force required}}{\text{force/tendon}} = \frac{975}{81·4} = 12$$

$$\therefore \text{spacing} = \frac{1000}{12} \quad \text{say} = 80\,mm.$$

Number of tendons per metre at top of tank

$$= \frac{225}{81·4} = 2·76.$$

$$\therefore \text{spacing} = \frac{1000}{2·76} \quad \text{say} = 280\,mm.$$

Intermediate spacings can be calculated from the diagram on page 6·3 (3).

Vertical Design
Tank empty.

When the tank is empty, moments will be induced in the vertical direction by the larger prestressing forces near to the foot of the wall, as compared with the smaller prestressing forces near to the top of the wall.

BS 5337
10·1.

The maximum moment induced may be assessed as being numerically equal to one half of the moment induced by a pinned base condition.

EXPLAN'TY.
HANDBOOK
BS. 5337
or
ACI TABLES.

Table A3.7.

$$\frac{h^2}{dt} = \frac{7·5^2}{20 \times 0·225} = 12·5.$$

Maximum value of coefficient = 0·0037.
Radial pressure due to prestressing

$$= \frac{\text{ring force}}{\text{radius}}$$

975 ←   ↓   → 975.

ref.	calculations	output
	EXAMPLE 6.3 SHEET 9.	

At top of tank, ring force   = 225 kN/m.

At bottom of tank, ring force   = 975 kN/m

∴ Radial pressure at top   = $\frac{225}{10}$ = 22.5

Radial pressure at bottom   = $\frac{975}{10}$ = 97.5.

∴ Assume uniform load of 22.5 kN/m² with a triangular load of 75 kN/m².

$q = 22.5$ kN/m²

$W_g = \frac{75}{7.5} = 10$ kN/m³

$M = K(W_g h^3 + q h^2)$

∴ Moment (hinged condition)

$= 0.0037 (10 \times 7.5^3 + 22.5 \times 7.5^2)$

$= 20.3$ kNm/m.

The design moment to be used is one half of 20.3 = 10.2 kNm/m. (tension on the outside).

Partially prestressed condition:

During the prestressing operations the tank will be compressed non-uniformly at each level.

The vertical stress produced may be estimated as numerically equal to 0.3 × the ring compressive stress at transfer.

ie. 0.3 × 4.8 = 1.44 N/mm².

Moment $= f_Z = \frac{1.44 \times 10^3 \times 225^2}{6 \times 10^6}$

$= 12.1$ kNm/m.

BS 5337
10.1.

ref.	calculations EXAMPLE 6·3 SHEET 10.	output
	This moment can be assumed to act in causing tension on both faces.	
	Section through tank when partially prestressed.	
	<u>Design of Reinforcement.</u>	
	Use normal reinforcement (ie. not prestressing).	
BS 5337 4.11.2.	Minimum reinforcement may be assumed to be 0.3%, as there is no restraint in the vertical direction except for friction at the base which causes some tension in the partially cast condition.	
	Equally horizontal reinforcement is necessary to control cracking before prestressing, and is used to support the vertical steel.	
	Horizontal reinforcement $$0.3\% \times 225 \times 1000 = 675 \text{ mm}^2.$$ <u>Use Y10 at 200 each face</u> (785).	Y10 at 200.

ref.	calculations EXAMPLE 6·3 SHEET 11.	output
TABLE A2.1. A2.2.	Vertical reinforcement. Total moment to be resisted $\qquad = 10·2 + 12·1 = 22·3$ kNm/m. Exposure class B. Table A2.1. $\qquad h = 200$ Y16 at 250. $M = 20·0$ $\qquad\qquad f_s = 200$ Table A2.2. $\qquad h = 250$ Y16 at 250 $M = 31·0$ $\qquad\qquad f_s = 225$. $\therefore$ at $f_s = 200$, $M = 31·0 \times \frac{200}{225}$ $\qquad\qquad\qquad = 27·5$. $\therefore$ approx value of moment of resistance. for $h = 225$ and 52 cover. $\qquad\qquad = \frac{1}{2}(20 + 27·5) = 23·7$ kNm. This is satisfactory. <u>Use Y16 at 250</u> vertically each face. If there are a number of tanks to construct, it may be economical to use welded wire fabric, which can be specially fabricated to provide the steel arrangements and sheet sizes required.	Y16 at 250

ref.	calculations. EXAMPLE 6·3 SHEET 12.	output

REINFORCEMENT DETAILS.

prestressing strands.

Y10 at 250 each face.

Y10 at 200 each face.

sliding joint.

7 Testing and Rectification

7.1 Testing for watertightness

The design and construction of liquid-retaining structures require close attention to detail by both the designer and contractor but, in spite of the best intentions of both parties, errors and omissions can occur. Equally, although the design theory outlined in this book has been used successfully for many structures, random occurrences and unfavourable statistical conjunctions can result in a structure which is less than completely liquid-tight. It is therefore necessary to test the structure after completion, to ensure that it is satisfactory and that it complies with the specification.

The method of test depends on the visibility and shape of the elements of the structure. The walls of overground structures can be inspected for leaks on the outer face and, if the walls are finally to be covered with soil, the inspection can be made before the fill is placed. The walls of underground structures can be inspected if there is sufficient working space available. The floor slabs of all structures built on soil cannot be inspected for leaks, and other methods of test have to be used[43]. The floor of an elevated reservoir (or water tower) can be inspected in the same way as walls, as can the underside of a flat reservoir roof. The detailed methods of testing are described in the following sections.

7.2 Water tests

A completed structure may be tested by filling with water and measuring the level over a period of time. The concrete in the structure must be allowed to attain its design strength before testing commences, and then the structure is slowly filled to its normal maximum operating level. If the structure is filled too quickly, the sudden increase in pressure is likely to cause cracking. As a guide, a swimming pool or relatively small tank could be filled over a period of three days, but a large reservoir will take much longer to fill because of the volume of water required. All outlets must be sealed to prevent loss of water through pipes, overflows and other connections.

To allow the concrete to become completely saturated with water, it is left for seven days after filling has been completed. The level of the water is then recorded, and subsequently each day for a further period of seven days. The difference in level over the period of seven days is then used to assess the result of the test. The levels may be measured by fixing scales to the walls, or by making marks on the walls above water line and measuring down to water level with a moveable scale or other device. The level should be recorded at four positions, but with a large reservoir at eight to twelve positions, to guard against errors in reading and local settlements.

An open reservoir (or a closed reservoir where the air above the water is affected by wind movements) may lose moisture by evaporation, or may gain water due to rainfall. In assessing the results of the water level readings during the test, allowance must be made for these variations. A simple method of achieving this is to moor a watertight container 80% filled with water at four points on the water surface. The water surface inside the container is subject to the same gains and losses as the water in the main reservoir. By taking measurements of the water level in the container from the top edge of the container, the gains or losses due to rainfall and evaporation in the main reservoir may be assessed (figure 7.1).

It will be apparent that a degree of honesty and care is necessary when carrying out tests of this nature, and the daily measurement of water levels during the test will assist in detecting any unusual occurrences.

7.3 Acceptance

A water test will enable a net loss of water to be measured due to leakage and further absorption into the concrete structure. The acceptable fall in water level should be stipulated by the designer before the test is commenced. For most structures, the maximum acceptable limit may be taken as 1/1000 of the average depth of the water. It is not possible to set a limit less than about 3 mm due to the difficulty of making a sufficiently accurate measurement.

If the test is judged to be unsatisfactory after seven days, and if the daily readings indicate that the rate of loss of water is reducing, the designer may

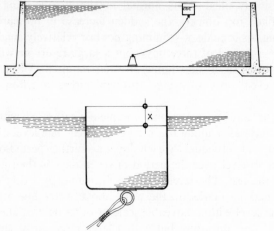

Figure 7.1   Arrangements for water test

decide to extend the test period by a further seven days. If the net loss of water is then no greater than the specified value during the second period of seven days, the test may be considered satisfactory.

If leakage from the external surface of the walls is observed, or if damp patches appear, they should be marked for future treatment. The structure should not be finally accepted as satisfactory until all apparent leaks or damp patches have been eliminated.

7.4   Remedial treatment

Small leaks and damp patches are usually self-healing after two to three weeks. After the healing is complete, accretions on the outside of the leak may be scraped off the surface. More persistent leaks require treatment with proprietary products, preferably from the water face. Chemicals are available which are applied to cracks as a slurry and are drawn into the crack by the water flow. Fine crystals are formed which close the crack. A similar effect occurs when the slurry is applied to a porous area. Areas of severe honeycombing or wide cracks may be repaired with pressure grouting techniques or, if there is severe leakage, a whole section may need to be cut out and replaced.

It is particularly difficult to isolate defects in floor slabs, but joints and areas where the concrete surface is irregular or honeycombed should be inspected very thoroughly. When water has been drained out of a structure, and the surface is drying, areas containing defects may be the last parts to remain wet or damp, due to water being trapped in the defective area.

Appendix A

A1.1 Limit state design

The two limit states that have a predominant effect on the thickness of concrete required and the quantity of reinforcement are:

(1) The limit state of cracking
(2) The ultimate limit state for flexure.

Generally (1) is the over-riding criterion. The first assessment of thickness of concrete section and reinforcement quantities for a preliminary design, in order to proceed to check that the two limit states are satisfied, is not easily possible by direct methods. Tables have therefore been prepared which enable the designer to choose sections that will resist a stated bending moment, in the knowledge that the chosen section will meet the requirements of both the limit states given above. An outline of the derivation of the Tables and the method of use follows.

A1.2 Layout of the design tables

The tables may be used to arrive at the overall thickness and the amount of reinforcement required to satisfy the ultimate limit state and the limit state of cracking. Values are tabulated for two limiting crack widths.

Tables A1.1—10 Class A exposure: limiting crack width = 0.1 mm
Tables A2.1—10 Class B exposure: limiting crack width = 0.2 mm

The layout of each separate table is similar. All the tables have been prepared for material properties and other conditions which are described in paragraph A1.5.4. Each is constructed for a particular overall section thickness h, varying between 200 mm and 1000 mm. Values of bending moment and reinforcement stress are given for standard bar sizes placed at standard spacings.

A1.3   Tabulated values

The values (in kN m) of bending moments which are given are the lesser of:
(1) The ultimate moment of resistance of the section divided by 1.6.
(2) The elastic moment of resistance of the section at the steel stress indicated, which together ensure that the crack width is within the limiting value for the particular class of exposure.

The formulae which have been used and the assumptions which have been made are given in paragraph A1.5.

A1.4   Method of using the tables

The tables should be used as follows:
(1) Calculate the applied service moment, i.e. the bending moment on the section due to the design loads (with a partial safety factor $\gamma_f = 1.0$).
(2) Decide on the exposure conditions and refer to the appropriate set of tables.
(3) Examine the tables for values of moment of resistance which are at least as great as the applied design moment, and decide on the thickness of the section. The precise size and spacing of reinforcement can be read, together with the design (service) stress in the reinforcement.
(4) There are several possible values of h and reinforcement quantity for any given value of applied moment, and the designer should be guided in his final choice by adopting a thickness h which shows the value of required design bending moment associated with the design tensile reinforcement stress desired. When deciding on an appropriate value for the thickness h of a section, it is also necessary to consider shear stresses. If no shear reinforcement is to be used, the ultimate shear stress must be limited.

A1.5   Derivation of tables

A1.5.1   *Computer program*
The values of bending moment and steel stress given in the tables have been calculated using an Olivetti P652 micro computer, and a program which accepts values of the parameters and calculates the width of cracks for a section with an initial arbitrary steel stress of 140 N/mm^2(MPa). The program then tests the crack width obtained and compares it with the permissible value. The stress is then increased or decreased by 5 N/mm^2(MPa) and the crack width recalculated. The program loops until the crack width is just less than the permissible value using steel stress

increments of $5\,N/mm^2$ (MPa). In view of the nature of the calculations, this degree of accuracy is acceptable. After the final calculation, the ultimate moment is calculated by limit state theory and divided by a partial safety factor for loads of 1.6. The factored ultimate moment and the service moment at the allowable crack width are then compared, and the lower value printed out. The value of the stress in the tensile reinforcement at the limiting value of crack width is also printed.

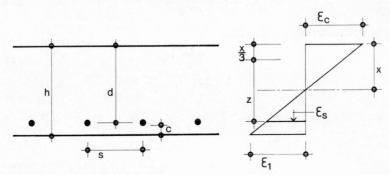

Figure A1.1

A1.5.2 *Formulae—elastic calculation*

The calculations are made by normal elastic theory for a unit length of slab reinforced in one face. Figure A1.1 illustrates a section through the slab. The following formulae are used in sequence to evaluate the properties of the section.

$$\text{Depth of neutral axis} = \frac{x}{d} = -\alpha_e\rho + \sqrt{(\alpha_e\rho)^2 + 2\alpha_e\rho} \quad \text{where} \quad \rho = \frac{A_s}{bd} \tag{A1.1}$$

$$\text{Lever arm } z = d - \tfrac{1}{3}x \tag{A1.2}$$

$$\text{Applied elastic moment } M = A_s f_{st} z \tag{A1.3}$$

Value of strain at surface ignoring stiffening effect of concrete

$$\varepsilon_1 = \frac{f_{st}}{E_s} \frac{h-x}{d-x} \tag{A1.4}$$

and, allowing for the uncracked concrete,

$$\varepsilon_m = \varepsilon_1 - \frac{0.7 b_t h (a' - x)}{A_s(h-x)f_{st}} \times 10^{-3} \tag{A1.5}$$

where f_{st} is in units of N/mm^2.

For cracking in a slab the factor

$$\frac{a'-x}{h-x}$$

in the formula is equal to unity.

Also A_s/bd is equal to the tensile steel ratio ρ and so

$$\varepsilon_m = \varepsilon_1 - \frac{0.7h}{\rho f_{st}d} \times 10^{-3} \qquad (A1.6)$$

The design surface crack width

$$w = \frac{4.5a_{cr}\varepsilon_m}{1+2.5\left(\dfrac{a_{cr}-c_{min}}{h-x}\right)} \qquad (A1.7)$$

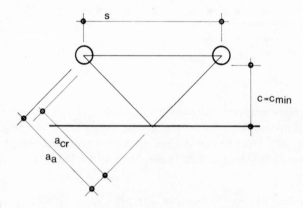

Figure A1.2

From figure A1.2 it can be seen that

$$a_{cr} = a_a - \tfrac{1}{2}\phi \qquad (A1.8)$$

$$a_a^2 = (s/2)^2 + (c+\tfrac{1}{2}\phi)^2 \qquad (A1.9)$$

Hence a_{cr} can be calculated and used in formula A1.7.

A1.5.3 Formulae—limit-state calculation
The ultimate lever arm is calculated

$$z = \left(1 - \frac{1.1f_yA_s}{f_{cu}bd}\right)d \qquad (A1.10)$$

and

$$M_u = (0.87f_y)A_sz \qquad (A1.11)$$

The factored value is $M_u/1.6$ and this value is printed in the tables if it is less than the service (or design) moment of resistance M_d obtained from the crack width calculation.

A1.5.4   Constants and material properties

The following values have been used in the computer program from which the tables have been prepared:

$$\text{Concrete: } f_{cu} = 25 \, \text{N/mm}^2 \, (\text{MPa})$$

$$\text{Reinforcement: high yield } f_y = 425 \, \text{N/mm}^2 \, (\text{MPa})$$

$$\text{Modular ratio } \alpha_e = 15$$

$$\text{Cover to tensile reinforcement } c = 52 \, \text{mm} \; (h = 200 \text{ to } 350)$$

$$60 \, \text{mm} \; (h = 400 \text{ to } 1000)$$

(It is assumed that the cover to the secondary reinforcement in the outer layer will be 40 mm.)

Area of tensile reinforcement is calculated from the specified bar sizes and spacings.

Area of compression reinforcement is assumed to be zero.

A1.5.5   Restrictions on tabulated values

1   If the arrangement of a particular bar size and spacing amounts to a steel ratio of less than 0.125% per face, no values of moment or stress are shown in the tables. This steel ratio is a practical minimum. Where values are tabulated, it does not follow that the particular steel ratio is necessarily sufficient to control early thermal movement, and this point must always be checked.

2   If a particular bar size and spacing, associated with a particular section thickness, results in a design concrete stress of more than $0.45f_{cu}$, no values are printed.

3   It should be noted that satisfactory designs may be prepared with values of moment and stress that are less than those shown in the tables. The tabulated values are absolute maxima derived from the formulae stated above. The judgement of the designer should be exercised in determining the reinforcement stress to be used at design loads.

A1.6   Example of use of tables A1 and A2

Design a cantilever wall 6.5 metres high to support a load due to water pressure. Class B exposure.

From table 3.1, for $H = 6.5\,\text{m}$ and $\dfrac{100A_s}{bd} = 0.5$, minimum thickness $h = 750\,\text{mm}$.

The values shown in tables A1 and A2 for the service moment of resistance of the section and the service stress in the reinforcement may be adjusted by applying the ratio to the steel stress of actual moment/tabulated moment. This is obviously only possible when the tabulated moment being considered is greater than the applied service moment (service conditions are with $\gamma_f = 1.0$).

Consider a unit width of slab of 1 metre:

Service moment $= M_s = (1/6) \times 10 \times 6.5^3 = 458\,\text{kN m/m}$

From table A2.9, for $h = 800\,\text{mm}$ consider Y25 at 150 ($3270\,\text{mm}^2/\text{m}$).

$$\text{Tabulated values: } M_s = 481\,\text{kN m/m}$$

$$f_s = 225\,\text{MPa}.$$

For

$$M = 458, \quad f_s = 225 \times \frac{458}{481} = 213\,\text{N/mm}^2$$

which is satisfactory.

∴ Use a slab with $h = 800\,\text{mm}$ and reinforced with Y25 at 150.

Using the values obtained from the design tables, a check on the accuracy of the original assumptions must now be made.

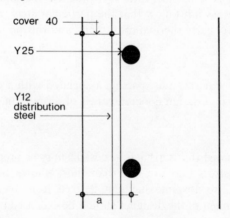

Figure A1.3

$$\text{Actual axial distance} = 40 + 12 + 12.5 = 65\,\text{mm}$$
$$d = 800 - 65 = 735\,\text{mm}$$
$$\text{Actual } \frac{100A_s}{bd} = \frac{100 \times 3270}{1000 \times 735} = 0.44$$

This is slightly less than the value of 0.50 which was assumed when checking the shear stress, and therefore the minimum thickness of 750 mm from table 3.1 will need to be increased. The value of 800 will be satisfactory.

Tables A1   Limiting moments (kN m) and steel stresses (N/mm^2) for 0.1 mm crack widths.

Table A1.1

$h = 200$ $c = 52$ Crack width $= 0.1$

Bar size (mm)	Bar spacing (mm)				
	100	150	200	250	300
12	20.3	16.7	14.5	13.2	
	145	175	200	225	
16		21.9	18.5	17.0	15.5
		135	150	170	185
20			23.1	20.3	18.3
			125	135	145
25				22.6	
				120	
32					

Table A1.2

$h = 250$ $c = 52$ Crack width $= 0.1$

Bar size (mm)	Bar spacing (mm)				
	100	150	200	250	300
12	33.7	26.8			
	175	205			
16	42.6	34.7	30.6	26.8	24.9
	130	155	180	195	215
20		43.8	37.2	33.3	29.7
		130	145	160	170
25			44.3	40.6	36.8
			115	130	140
32					

Table A1.3

$h = 300$ $c = 52$ Crack width = 0.1

Bar size (mm)	Bar spacing (mm)				
	100	150	200	250	300
12	47.8 195	38.3 230			
16	62.9 150	50.0 175	43.5 200	38.6 220	
20	79.1 125	62.7 145	52.7 160	48.0 180	42.6 190
25			64.5 130	58.3 145	52.5 155
32					66.3 125

Table A1.4

$h = 350$ $c = 52$ Crack width = 0.1

Bar size (mm)	Bar spacing (mm)				
	100	150	200	250	300
12	64.2 215				
16	84.5 165	68.0 195	58.3 220		
20	104.6 135	84.6 160	70.4 175	63.5 195	57.4 210
25		103.7 130	88.2 145	78.9 160	70.5 170
32				96.8 125	88.1 135

Table A1.5

$h = 400$ $c = 60$ Crack width = 0.1

Bar size (mm)	Bar spacing (mm)				
	100	150	200	250	300
12	75.6 220				
16	100.4 170	82.3 205	70.1 230		
20	125.3 140	100.7 165	85.9 185	76.9 205	69.3 220
25		124.5 135	105.5 150	94.0 165	86.2 180
32			132.8 120	116.7 130	105.8 140

Table A1.6

$h = 450$ $c = 60$ Crack width = 0.1

Bar size (mm)	Bar spacing (mm)				
	100	150	200	250	300
12					
16	126.6 185	100.0 215			
20	155.8 150	123.9 175	107.6 200	95.6 220	
25	196.2 125	155.5 145	130.7 160	115.7 175	105.6 190
32			167.5 130	146.3 140	131.9 150

Table A1.7

$h = 500$ $c = 60$ Crack width = 0.1

Bar size (mm)	Bar spacing (mm)				
	100	150	200	250	300
12					
16	151.9 195	121.5 230			
20	189.4 160	149.1 185	128.6 210	113.7 230	
25	232.8 130	189.5 155	158.3 170	143.1 190	129.7 205
32		231.4 120	198.7 135	178.9 150	160.6 160

Table A1.8

$h = 600$ $c = 60$ Crack width = 0.1

Bar size (mm)	Bar spacing (mm)				
	100	150	200	250	300
12					
16	208.1 215				
20	258.2 175	205.6 205			
25	324.5 145	259.4 170	220.5 190	197.0 210	177.2 225
32		326.0 135	276.2 150	246.0 165	225.7 180

Table A1.9

| | $h = 800$ | $c = 60$ | Crack width $= 0.1$ | | |

Bar size (mm)	Bar spacing (mm)				
	100	150	200	250	300
12					
16					
20	423.0 205				
25	534.2 170	416.9 195	357.3 220		
32	672.3 135	526.8 155	452.9 175	397.6 190	360.4 205

Table A1.10

| | $h = 1000$ | $c = 60$ | Crack width $= 0.1$ | | |

Bar size (mm)	Bar spacing (mm)				
	100	150	200	250	300
12					
16					
20	611.1 230				
25	770.8 190	606.2 220			
32	967.2 150	769.0 175	651.6 195	580.5 215	521.4 230

Tables A2   Limiting moments (kN m) and steel stresses (N/mm^2) for 0.2 mm crack widths.

Table A2.1

$h = 200$ $c = 52$ Crack width = 0.2

Bar size (mm)	Bar spacing (mm)				
	100	150	200	250	300
12	19.6 140	19.6 205	16.7 230		
16		26.7 165	22.2 180	20.0 200	17.6 210
20			28.7 155	24.8 165	22.1 175
25					27.4 145
32					

Table A2.2

$h = 250$ $c = 52$ Crack width = 0.2

Bar size (mm)	Bar spacing (mm)				
	100	150	200	250	300
12	39.5 205				
16	54.1 165	42.5 190	35.7 210	31.0 225	
20		53.9 160	44.9 175	39.5 190	35.0 200
25			57.8 150	50.0 160	43.4 165
32					

Table A2.3

$h = 300$ $c = 52$ Crack width $= 0.2$

Bar size (mm)	Bar spacing (mm)				
	100	150	200	250	300
12	56.4 230				
16	77.6 185	60.0 210	50.0 230		
20	101.2 160	75.6 175	62.6 190	54.6 205	49.3 220
25			81.9 165	70.4 175	62.6 185
32					79.6 150

Table A2.4

$h = 350$ $c = 52$ Crack width $= 0.2$

Bar size (mm)	Bar spacing (mm)				
	100	150	200	250	300
12					
16	102.5 200	78.4 225			
20	135.6 175	100.4 190	82.5 205	71.6 220	
25		131.6 165	106.5 175	91.2 185	80.8 195
32				120.1 155	104.4 160

Table A2.5

$h = 400$ $c = 60$ Crack width $= 0.2$

Bar size (mm)	Bar spacing (mm)				
	100	150	200	250	300
12					
16	118.2 200	92.4 230			
20	152.1 170	119.0 195	97.5 210	86.3 230	
25		152.2 165	123.1 175	108.2 190	95.8 200
32			160.5 145	139.1 155	124.7 165

Table A2.6

$h = 450$ $c = 60$ Crack width $= 0.2$

Bar size (mm)	Bar spacing (mm)				
	100	150	200	250	300
12					
16	147.2 215				
20	192.2 185	145.1 205	121.1 225		
25	243.3 155	187.6 175	155.2 190	132.2 200	119.5 215
32			199.8 155	172.4 165	153.9 175

Table A2.7

$h = 500$ $c = 60$ Crack width = 0.2

Bar size (mm)	Bar spacing (mm)				
	100	150	200	250	300
12					
16	175.2 225				
20	224.9 190	173.2 215			
25	295.5 165	220.1 180	186.2 200	158.1 210	142.4 225
32		299.0 155	242.9 165	208.7 175	185.7 185

Table A2.8

$h = 600$ $c = 60$ Crack width = 0.2

Bar size (mm)	Bar spacing (mm)				
	100	150	200	250	300
12					
16					
20	309.8 210				
25	402.9 180	305.2 200	249.6 215	215.8 230	
32		398.4 165	331.5 180	283.3 190	250.8 200

Table A2.9

$h = 800$	$c = 60$	Crack width = 0.2

Bar size (mm)	Bar spacing (mm)				
	100	150	200	250	300
12					
16					
20					
25	644.2 205	481.0 225			
32	846.6 170	645.8 190	517.6 200	449.9 215	395.6 225

Table A2.10

$h = 1000$	$c = 60$	Crack width = 0.2

Bar size (mm)	Bar spacing (mm)				
	100	150	200	250	300
12					
16					
20					
25	892.5 220				
32	1225.2 190	900.8 205	735.2 220		

Appendix B

Rectangular slab panels: 2-way spans

Figure B.1 enables slab panels to be designed when loaded with triangularly distributed loads. An additional surcharge pressure with rectangular distribution can be designed by reference to table 17 in CP 110, or by considering the wall to be extended by a height equal to (surcharge/density of liquid).

An example of the use of figure B.1 is given in Example 6.1.

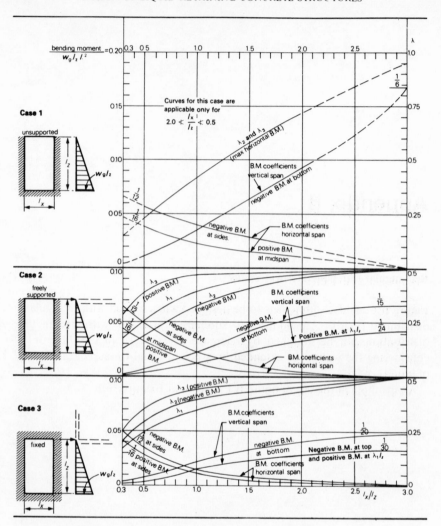

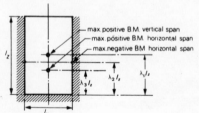

$l = l_x$ or l_z as appropriate

Appendix C

Bar reinforcement

The tables of bar areas in table C.1 are included for convenience. The spacings and bar sizes are the preferred values in UK.

  Abbreviations: (UK practice)

  Y = High yield deformed bars

  R = Mild steel plain bars

Figure B.1 Two-way slabs: rectangular panels, triangularly-distributed loads
  Panels fixed or continuous along bottom edge and both vertical sides; condition along top edge as indicated.
  Fractions thus '$\frac{1}{12}$' indicate coefficients to which curves are asymptotic or to which coefficients approach as span ratio l_x/l_z approaches zero or infinity.

Vertical span: bending moment = (coefficient) fl_z^2

Horizontal span: bending moment = (coefficient) fl_x^2

Scale on right-hand side is for values of λ_1, λ_2 and λ_3. Ratio of spans = $k = l_x/l_z$.

Table C.1

Sectional Areas of Groups of Bars (mm²)

Bar size (mm)	Number of bars									
	1	2	3	4	5	6	7	8	9	10
6	28.3	56.6	84.9	113	142	170	198	226	255	283
8	50.3	101	151	201	252	302	352	402	453	503
10	78.5	157	236	314	393	471	550	628	707	785
12	113	226	339	452	566	679	792	905	1020	1130
16	201	402	603	804	1010	1210	1410	1610	1810	2010
20	314	628	943	1260	1570	1890	2200	2510	2830	3140
25	491	982	1470	1960	2450	2950	3440	3930	4420	4910
32	804	1610	2410	3220	4020	4830	5630	6430	7240	8040
40	1260	2510	3770	5030	6280	7540	8800	10100	11300	12600

Sectional Areas per metre width for various Bar Spacings (mm²)

Bar size (mm)	Spacing of bars (millimetres)								
	50	75	100	125	150	175	200	250	300
6	566	377	283	226	189	162	142	113	94.3
8	1010	671	503	402	335	287	252	201	168
10	1570	1050	785	628	523	449	393	314	262
12	2260	1510	1130	905	754	646	566	452	377
16	4020	2680	2010	1610	1340	1150	1010	804	670
20	6280	4190	3140	2510	2090	1800	1570	1260	1050
25	9820	6550	4910	3930	3270	2810	2450	1960	1640
32	16100	10700	8040	6430	5360	4600	4020	3220	2680
40	25100	16800	12600	10100	8380	7180	6280	5030	4190

Note: The above tables have been calculated to three significant figures according to BSI recommendations.

Bibliography

British Standards

The British Standards Institution publications referred to in this book, or of relevance to the subject are listed below:

BS 12   Portland cement (ordinary and rapid-hardening): Part 2: Metric units
BS 882, 1201   Aggregates from natural sources for concrete (including granolithic)
BS 1370   Low heat Portland-blastfurnace cement
BS 2499   Hot applied joint sealants for concrete pavements
BS 4027   Sulphate-resisting Portland cement: Part 2: Metric units
BS 4246   Low heat Portland-blastfurnace cement
BS 4254   Two-part polysulphide-based sealants for the building industry
BS 4449   Hot rolled steel bars for the reinforcement of concrete
BS 4461   Cold worked steel bars for the reinforcement of concrete
BS 4483   Steel fabric for the reinforcement of concrete
BS 5337   The structural use of concrete for retaining aqueous liquids
BS 5896   High tensile wire and strand for the prestressing of concrete
CP 110   The structural use of concrete
               Part 1: Design, materials and workmanship
               Part 2: Design charts for singly reinforced beams, doubly reinforced beams and rectangular columns
               Part 3: Design charts for circular columns and prestressed beams
CP 114   Structural use of reinforced concrete in buildings: Part 2: Metric units
CP 115   Structural use of prestressed concrete in buildings: Part 2: Metric units
CP 2004   Foundations

American Standards and Codes of Practice

American Society for Testing and Materials

A 82—72   Specification for cold-drawn steel wire for concrete reinforcement
A 185—73   Specification for welded steel wire fabric for concrete reinforcement

A 416—68	Specification for uncoated seven-wire stress-relieved strand for prestressed concrete
A 421—65 (1972)	Specification for uncoated stress-relieved wire for prestressed concrete
A 496—72	Specification for deformed steel wire for concrete reinforcement
A 497—72	Specification for welded deformed steel wire fabric for concrete reinforcement
C 33—74	Specification for concrete aggregates
C 39—72	Method of test for compression strength of cylindrical specimens
C 150—74	Specification for Portland cement
C 494—71	Specification for chemical admixtures for concrete
D 994—71	Specification for preformed expansion joint filler for concrete (Bituminous type)
1190—64 (1970)	Specification for concrete joint sealer, hot-poured elastic type
1850—67 (1972)	Specification for concrete joint sealer, cold-application type

American Concrete Institute

AC I 301—72 (revised 1973)	Specification for structural concrete for buildings
AC I 308–71	Recommended practice for curing concrete
AC I 309—72	Recommended practice for consolidation of concrete
AC I 318–77	Building code requirements for reinforced concrete
AC I Committee 515 Report (63–59)	Guide for the protection of concrete against chemical attack by means of coatings and other corrosion resistant materials

References

(1) Closner, J. J. and Porat. M. M., 'Leakproof concrete tanks for aviation fuel,' *Civil Engineering*, USA, April 1960.
(2) Manning, G. P., *Reinforced Concrete Reservoirs and Tanks*, Cement and Concrete Association, London, 1972.
(3) ACI Committee 515, 'Guide for the protection of concrete against chemical attack by means of coatings and other corrosion-resistant materials,' *Journal of the American Concrete Institute. Proceedings* Vol. 63, No. 12, December 1966, pp. 1305–1392.
(4) Gutt, W. H. and Harrison, W. H., 'Chemical resistance of concrete,' Building Research Current Paper CP 23/77, Building Research Station, Garston, 1977.
(5) ACI 318–77, *Building Code Requirements for Reinforced Concrete*, American Concrete Institute, Detroit, 1977.
(6) CP 110, *The Structural Use of Concrete, Part 1 : Design, Materials and Workmanship*, British Standards Institution, 1972.
(7) Comité Européen du Béton—Fédération Internationale de la Précontrainte, *International Recommendations for the Design and Construction of Concrete Structures*, English edition, Cement and Concrete Association, London, 1970.
(8) Base, G. D., Read, J. B., Beeby, A. W. and Taylor, H. P. J., 'An investigation of the crack control characteristics of various types of bar in reinforced concrete beams,' Research Report 41-018, Cement and Concrete Association, London, 1966.
(9) Hughes, B. P., *Limit State Theory for Reinforced Concrete Design*, Pitman, second edition, 1976.
(10) BS 5337: *Code of Practice for the Structural Use of Concrete for retaining Aqueous Liquids* (formerly CP 2007), British Standards Institute, 1976.
(11) ACI 350 R-77, Report: *Concrete Sanitary Engineering Structures*, American Concrete Institute, 1977.
(12) Sadgrove, B. M., *Water Retention Tests of Horizontal Joints in Thick-walled Reinforced Concrete Structures*, Cement and Concrete Association, London, 1974.

(13) Davies, J. D., 'Circular tanks on ground subjected to mining subsidence,' *Civil Engineering and Public Works Review*, Vol. 55, July 1960, pp. 918–920.

(14) Gray, W. S. and Manning, G. P., *Concrete Water Towers, Bunkers, Silos and other elevated Structures*, Cement and Concrete Association, London, 1973.

(15) 'Protection against corrosion of reinforcing steel in concrete,' Building Research Station Digest No. 59, HMSO, London, 1965.

(16) Teychenne, D. C., Franklin, R. E. and Erntroy, H. C., *Design of Normal Concrete Mixes*, HMSO, London, 1975.

(17) Palmer, D., *Concrete Mixes for General Purposes*, Cement and Concrete Association, London, 1977.

(18) Bannister, J. L., 'Steel reinforcement and tendons for structural concrete,' *The Consulting Engineer*, Vol. 35, No. 2, February 1971.

(19) BS 4449: *Hot Rolled Bars for the Reinforcement of Concrete*, British Standards Institution, 1978.

(20) BS 4461: *Cold Worked Bars for the Reinforcement of Concrete*, British Standards Institution, 1978.

(21) BS 5896: *High-tensile Wire and Strand for the Prestressing of Concrete*, British Standards Institution, 1980.

(22) 'Shrinkage of natural aggregates in concrete,' Building Research Station Digest No. 35 (Second series), HMSO, London, 1968.

(23) CP 2004, *Code of Practice for Foundations*, British Standards Institution, 1972.

(24) Wilun, Z. and Starzewski, K., *Soil Mechanics in Foundation Engineering*, Vol 2, *Theory and Practice*, Surrey University Press, Second edition, 1975.

(25) 'Mini-leap computer program,' Computer Consortium Services Ltd., London.

(26) Anchor, R. D., Hill, A. W. and Hughes, B. P., *Handbook to BS 5337*, Viewpoint Publication No. 14-011, Cement and Concrete Association, London, 1979.

(27) 'Standard method of detailing reinforced concrete,' Report of a Joint Committee drawn from the Concrete Society and the Institution of Structural Engineers, 1970.

(28) 'Standard reinforced concrete details,' Concrete Society, 1973.

(29) Higgins, J. B. and Hollington, M. R., *Designed and Detailed*, Cement and Concrete Association, London, 1973.

(30) Bate, S. C. C. and Bennett, E. W., *Design of Prestressed Concrete*, Surrey University Press, 1976.

(31) Cahill, T. and Branch, G. D., 'Long-term relaxation behaviour of stabilized prestressing wires and strands,' Paper No. 19, Conference on prestressed concrete vessels, London, March 1967, Institution of Civil Engineers, London, 1968.

(32) Creasy, L. R., *Prestressed Concrete Cylindrical Tanks*, pp. 216, Contractors Record Limited, London, 1961.

(33) American Concrete Institute, 'Design and construction of circular prestressed concrete structures,' Title 67-40, *J. ACI* 6(9), September 1970, pp. 657–672.

(34) Evans, E. P. and Hughes, B. P., 'Shrinkage and thermal cracking in a reinforced concrete retaining wall,' *Proceedings of the Institution of Civil Engineers*, Vol. 39, January 1968, pp. 111–25.

(35) Lea, F. M., *The Chemistry of Cement and Concrete*, Edward Arnold, London, third edition, 1970.

(36) Fitzgibbon, M. E., 'Thermal controls for large pours,' *Civil Engineering and Public Works Review*, September 1973.

(37) Fitzgibbon, M. E., *Continuous Casting of Concrete. New Concrete Technologies and Building Design*, ed. Neville, A. M., The Construction Press, 1979.

(38) Fitzgibbon, M. E., 'Large pours,' Current Practice Sheets Nos. 1–3, *Concrete*, March 1976, December 1976, February 1977.

(39) Priestley, M. J. N., 'Ambient thermal stresses in circular concrete tanks,' *J. ACI*. October 1976, pp. 553–60.

(40) Deacon, R. C., *Watertight Concrete Construction*. Cement and Concrete Association, London, 1978.

(41) Monks, W. L., 'The performance of waterstops in movement joints,' Technical Report No. 42-475, Cement and Concrete Association, London, 1972.
(42) ACI 318-63: Building code requirements for reinforced concrete. American Concrete Institute, Detroit, 1963.
(43) 'Civil Engineering Specification for the water industry,' National Water Council, London, 1978.
(44) 'Admixtures for concrete,' Technical report TRCS 1, The Concrete Society, London, December 1967.

General

Deacon, C. R., 'Concrete ground floors: their design, construction and finish,' Advisory publication 48-034. Cement and Concrete Association, London, 1974.

Green, J. K. and Perkins, P. H., *Concrete Liquid Retaining Structures—Design, Specification and Construction*, Applied Science Publishers, 1980, 355 pp.

Reinforcement Manufacturers' Association, 'Reinforced concrete ground slabs—a guide to their design,' April 1971, 9 pp.

Reynolds, C. E. and Steedman, J. C., *Reinforced Concrete Designers Handbook*, Viewpoint Publications, Cement and Concrete Association, London, 1976.

Threlfall, A. J., 'Design charts for water-retaining structures—to BS 5337,' Viewpoint Publication No. 12-078, Cement and Concrete Association, London, 1978.

Index